BIG BANG

BIG BANG

By Ian M. Kerr

Published by
Euromoney Publications
in association with
Alexanders Laing & Cruickshank Holdings Ltd
Barings plc
Dow Financial Services Ltd
Kleinwort, Benson Ltd
Macfarlanes
Midland Bank Treasury, Samuel Montagu & Co Ltd
and Greenwell Montagu & Co
Price Waterhouse
Shearson Lehman Brothers International, Inc

Published by
Euromoney Publications PLC
Nestor House, Playhouse Yard
London EC4V 5EX

ISBN 1 870031 30 X

Typeset by PW Graphics
Printed in England by
Balding & Mansell Ltd

Contents

Foreword

by Sir Kenneth Berrill

The admission of corporate members to the Stock Exchange, and all that has preceded it in the past two years, has already changed the face of the UK securities market almost beyond recognition. The effects yet to come of the actual events of 27 October 1986, the disappearance of minimum commissions on dealings and the official end of 'single capacity' for Stock Exchange member firms, are likely to be hardly less significant.

The ripples caused by the arrangements made in 1983, which resulted in the withdrawal of a case brought against the Stock Exchange in the Restrictive Practices Court by the Director-General of Fair Trading, have spread to all sectors of the industry, resulting in the spread of new conglomerates, in new methods of operation and in a surge of investment in new technology to cope with developments. And, on top of all this turbulence which is affecting most major securities markets worldwide, comes the imminent arrival of major new regulation through the provisions of the Financial Services Bill which was drawn up specifically to provide investor protection throughout all forms of investment business (other than property markets.)

As Chairman of the Securities and Investments Board (SIB), which is expected to be designated as the agency to regulate financial services under the legislation, I am well aware of the general feeling that the industry has more than enough change to cope with already, without further and changing regulations. The argument is quite understandable. But it is also a very short-term view at a time when long-term thinking is essential.

There is nothing intrinsically incompatible between the concepts of market efficiency and expansion and proper protection of investors. And, although the Stock Exchange has a superb record of regulating its activities, it is not the only market involved in the re-styled City. Some form of comprehensive regulation to match the nature of comprehensive investment conglomerates seems both logical and necessary, particularly since so much emphasis is now being placed on bringing small investors into more active participation as direct owners of shares, holders of complex life assurance policies and possibly speculators in futures and options markets.

Although some of the detail has been changed since the Bill was first published in 1985, its basic principles and intent remain unaltered. At the heart of the system is the requirement that anyone authorised to carry out investment business must be 'fit and proper' to do so. 'Fit and proper' meaning honest and competent, adequately capitalised, maintaining good records, keeping clients' money separate from firms' money, etc. From that requirement flows, inevitably, a great deal of regulatory detail. But the SIB has a large base of practitioner expertise on which to draw when exercising its duty to provide investor protection without unduly impeding the efficiency or competitiveness of the businesses concerned. It is fully aware of the need to treat professional operators in the market with a much lighter protective hand to that designed to help 'Aunt Agatha'. The industry cannot reasonably be expected to like the idea of new regulation, and that sense of discomfort is leading to an inevitable tendency on the part of a few to magnify all conceivable (and some remarkably improbable) problems out of all proportion.

But the experience of the United States, in both regulation of the securities market and regulation of the investment industry generally, does underline the lesson that well-regulated markets create such increased investor confidence that the volume of trading is thereby considerably increased, and the 'cost' of improved regulation is more than absorbed by the economies of greater through-put. What is happening now, both on the market and legislative fronts, should help to ensure that London keeps its place as a major, and growing, international financial centre.

Preface

The deregulation of the London capital markets, the dismantling of fixed commissions and the admission of foreign member firms signals the end of an era for the London Stock Exchange. There are many who will regret the passing of that era and soon the period prior to October 1986 will be seen with nostalgia as 'the good old days'. The harsh reality is that bygone times presented a serious obstacle to necessary change. The London Stock Exchange was becoming like a last outpost in a faraway colonial possession. But it was never under siege because any potential aggressors were deterred by its own insularity. If a relief column had been necessary, the City of London had to look no further than to the Euromarkets, which had arrived 20 years before, expanded and prospered through a combination of non-interference and perhaps just a little domestic ignorance, and now threatened to outgrow the London Stock Exchange.

Because the City of London has so many blueprint examples to follow, the changes within its financial markets will not cause any real problems. When the then Secretary of State for Trade and Industry Cecil Parkinson and the Chairman of the London Stock Exchange Sir Nicholas Goodison reached their agreement in the summer of 1983, the facts were relatively simple. In July 1983, the London Stock Exchange, even according to some of its most prestigious members, was probably 10 years out of date. The Parkinson–Goodison accord gave the Stock Exchange three and a quarter years to make the necessary structural adjustments which would ensure that it remained one of the three major financial centres of the world. The fact that it was never really in danger of losing its place among this trio is a reflection of the City of London's remarkable history as well as a considerable element of luck. The luck was the presence of the Euromarkets, which had essentially been created in London and had been all but ignored by the London investment community. With a few notable UK exceptions, the huge Euromarkets were gift-wrapped by the City of London's banks and stockbrokers and were handed to foreign financial institutions.

The example of the Euromarkets does not augur well for the survival of London's new financial consortia in a deregulated market, faced by wealthy foreign competitors with long experience and expertise in market making skills. The popular conception is that the UK houses will be overwhelmed by the overseas opposition. Those who support this theory would do well to look back in history and study the tenacity and creativity which has characterised the British financial institutions over many hundreds of years. It is a proud heritage which they will not surrender easily.

The road to deregulation has been an author's nightmare and should be sufficient to deter any aspiring writer to attempt the coverage of a subject which involves moving legislation. The Financial Services Bill, around which the deregulation of the City of London's financial system will be based, began as a basic document of 170 pages before additions. From the start it was easy to see that it was a report for practitioners prepared by non-practitioners. At the time of writing, amendments were still being tabled to cure the worst excesses. Because the final version of the Bill has not been concluded, authors and commentators can make no final judgements, but it is absolutely certain that not everyone will be pleased. The aftermath of deregulation will not be known for two or three years. In the immediate post-deregulation period a number of smaller firms will be forced to merge or to cease business. In the near future some highly respected household names in the medium-sized category will merge their operations because those which share 'the common dilemma of middle ground' may find that position untenable. This is not to say that size alone will be the only determinant of survival. Niche companies make handsome profits as the major houses compete for market share at the expense of profit. Some of the new financial consortia will find that they have paid too much for their subsidiaries. Personal working and cultural philosophies within certain consortia suggest substantial differences which will have to be resolved. City mergers made hundreds of new millionaires, but no-one has asked whether they are all happy with their lot.

As I firmly believe that the future London securities markets will move rapidly towards 'Euro' style trading, this may be interpreted as an assumed *fait accompli*. This may not prove to be the case and would upset some hard-line UK traditionalists. The suggested list of 'Division One' members of the gilt-edged markets will cause more concern among those who were omitted rather than those included. The rankings may cause early resentment, but the real outcome will not be known for a number of years. If there is some bias against the Stock Exchange, this should be considered only against the background of someone who has watched, from an independent vantage point, some magnificent missed opportunities.

There are numerous people whose counsel, help and sympathy with bureaucratic inertia have contributed signficantly towards the completion of this book. They include Cecil Parkinson, Charles McVeigh III, Sheldon Prentice, Michael Hawkes, Lord Rockley, David Benson, John Barkshire, Miles Rivett-Carnac, Hans-Joerg Rudloff, Richard Briance, Richard Bristow, Larry Woodell, Jacques Gelardin, Win Bischoff, Rupert Hambro, Stanislas Yassukovich, Ian Steers, Charles Villiers, David Stewart, Vanni Treves, Stuart Valentine, Lord Camoys, John Sanders, Stanley Ross, Paul Sherwood, Philip Hubbard, Michael Perry, Gordon Pepper, George Westropp, Sir Kenneth Berrill, Sir Jeremy Morse, Archibald Cox and David Poole. Among my colleagues within Kidder, Peabody & Co the impartial advice of Mohamed Younes and the encyclopaedic memory of Peter Stuart William are greatly appreciated. Cathy Casey, who typed the manuscript, displayed the speed and efficiency which makes her so valuable.

Ian M. Kerr
London, August 1986

The deregulation of the City – its impact on London's financial markets and institutions

Perhaps the most surprising aspect of the City of London's proposals for deregulation was that they took so long in coming. If ever a major financial centre required fundamental change it was London. The City was living on its past reputation. Tourists still flocked to see the bill brokers in their top hats, and guides pointed out the original locations of the coffee houses where business transactions first took place in the seventeenth century. Lloyds of London was still the insurance hub of the world. Some of the overseas tourist guides confused the Royal Exchange with the Stock Exchange. It did not really matter. The overall impression was one of orderly calm. Perhaps of even more importance was the impression of institutional wealth. In the eyes of visitors, banks represent money. The City of London could offer them a choice of 592 in 1986. Nor were they parochial. Out of the 592 banks and representative offices 399 were non-British. Against this background it was impossible for outsiders to imagine that they were not at the very centre of international finance.

However, cracks were already beginning to appear in the façade. Progression was painfully slow. The City became a follower rather than a leader. LIFFE (the London International Financial Futures Exchange) was established in September 1982. In the United States, financial futures had been traded on the Chicago Board of Trade since 1977. The New York Stock Exchange, as well as the other US stock exchanges, had abandoned fixed commissions on share transactions in May 1975. Wall Street had suffered considerable attrition among its member firms, but had recovered to achieve record levels of profitability. The London Stock Exchange was now following the Wall Street example more than 11 years later in its decision to eliminate fixed rate commissions on equity and bond transactions. Lloyds of London became embroiled in a succession of insurance scandals, whose effects were felt far beyond the City.

As London's reputation as a financial centre began to wane, it was not only the foreigners who began to ask questions. UK political leaders and Treasury mandarins began enquiring about efficiency, innovative abilities, the more competitive edge of the foreign opposition, a coherent strategy for the future and the level of investor protection for UK residents. All had a degree of justification. The London Stock Exchange was regarded by its overseas counterparts as 'a walking anachronism'. The criticism was certainly unfair, but no foreign members were permitted and it was therefore easy to qualify accusations of blatant protectionism and the existence of a cartel. Most stockbroking firms were woefully undercapitalised, as were the few remaining jobbing firms, in relation to their US or Japanese counterparts. Efficiency in the eyes of international investors came to be interpreted as lower dealing prices. The role of the single-capacity division between stockbrokers and jobbers was therefore examined to see if dual capacity, in the form of net market making, might not be a more preferable and cheaper alternative. That the Stock Exchange was perceived, however incorrectly, to be holding out against the abolition of fixed commissions only reinforced the insinuations of a cartel.

Innovative capabilities have always been encouraged and have generally thrived in the City. However, the Stock Exchange had not been at the forefront of changes, which became permanent components of the fund raising mechanism. Its technology was inadequate and the member firms generally lacked the capital resources to install the new systems.

The Stock Exchange's failure to recognise the potential of the fledgling Eurobond market must go down as one of the most extraordinary oversights in financial history. The internationally syndicated public Eurobond issue was pioneered in mid-1963 by the late Sir Siegmund Warburg, then Chairman of S. G. Warburg & Co, for one of the Italian state agencies. Although growth was slow in the initial stages, volume had reached a sufficient level by 1965 for the then small number of participating houses in the Eurobond market to approach the London jobbers with the request that they trade the new financial instruments. Julius Strauss, a partner of Strauss Turnbull & Co, and one of the inspirations behind the success of the Eurobond market, was among the delegation to the jobbers, which included Hambros Bank and S. G. Warburg & Co. In *A History of the Eurobond Market – the First 21 Years** he recalled 'they were simply not very interested'. The banks, too, made every effort to persuade the jobbers that the new market had a future. Although strict foreign exchange controls were in existence at the time, Hambros Bank offered to provide US dollar financing to match the jobbers' trading books, but talks never progressed beyond the exploratory stage.

The exact amount of money which this lack of recognition cost the London jobbing system is incalculable, but must run into tens of millions of dollars, which were sorely needed before exchange controls in the United Kingdom were removed in 1979. To add insult to injury, the UK jobbers then watched the Eurobond market, which included a high proportion of industrial borrowers, move inexorably towards new records, while their own UK domestic corporate sterling market almost closed completely between 1972 and 1980. It was only when Eurosterling and Bulldog issues became a real factor that jobbing houses such as Akroyd & Smithers, now part of the Mercury International group, began to play an active part in market making. As Julius Strauss recently pointed out, 'The jobbers' loss was our gain'.

By modern trading standards, where transactions are concluded in seconds, the early Euro trades involved layers of back-office processing. There was no central clearing system. Euroclear and Cedel did not emerge for another five years. Most transactions involved the physical delivery of bonds which is expensive, time-consuming, inefficient and insecure. Interest coupons were clipped by hand and again physically presented for payment. Because 'dual capacity' had not even been considered, firms such as Strauss Turnbull were not allowed to act as direct market makers and although they effectively acted as principals, the transactions were entered through the books of the London stock jobbers, irrespective of the fact that the jobbers did not make a bid or offer price. Single-capacity rules were strictly observed. In those early days, Stanley Ross, then of Strauss Turnbull & Co and now a Managing Director of Deutsche Bank Capital Markets in London, recalls quoting prices over the telex with the preface that 'We have buyers at a price of *x* and sellers at a price of *y*'. The stockbrokers' role as an agent was therefore preserved but, in effect, the early Eurobond transactions foresaw the elimination of the single-capacity system in the City of London more than 20 years before it was legally implemented.

The foreign invasion

Foreign competition in the City of London is not new. The most obvious presence is the substantial number of foreign banks with branch or representative offices. The majority came, not only to establish themselves in the United Kingdom, but to tap the huge pool of Eurodollars which had accumulated outside the United States after the Second World War. Their activities, particularly those of the new wave of US banks, were limited. Their knowledge of international markets and international lending was scant. 'We participated in the early Euro-syndicated loans because my Chairman wanted to tell his board that the bank had an established overseas presence', recalls one former banker from Wisconsin. 'To be frank', he added, 'the Chairman probably didn't even know where some of our borrowers were located.'

The Euro-syndicated loan market became the fiefdom of foreign banks, with British commercial bank participation only becoming a factor at a later stage. In the securities markets the foreigners were often more ambitious. While they could not participate directly in the London Stock Exchange, except through the medium of a London stockbroker, they quickly found a niche in the Eurosecurities markets, where membership was open to anyone with reasonable credentials. Therefore, the US house of White Weld & Co (now part of the Credit Suisse First Boston Group) became the leading secondary market maker in Eurobonds, as well as being a formidable force in the primary sector. Other US banks, such as Security Pacific, took the consortium route into the Euromarket via a holding in Western American Bank, which included powerful UK shareholders such as Hambros Bank Ltd. Although Western American Bank enjoyed a few highly profitable years in the late 1960s, it eventually closed its Euro operations in September 1974, one of the first casualties of an inverted yield curve.

However, the foreign institutions in London were still proving to be the more inventive. The first European Depository Receipts for Japanese companies in the mid-1960s, representing a specific number of underlying ordinary shares, were the forerunner of today's Euroequities. The UK institutional investors had long been adept at identifying profitable investment opportunities overseas, even when they had to pay a penal level of premium for the privilege of investing abroad. The failure of London's financial community to match its international portfolio selection expertise with similar skills in international distribution is an omission which will be exploited fully by overseas participants in London after deregulation.

Will the City of London be able to survive the competition or will it simply succumb to being a satellite of US and Japanese financial ambition? For a time there will be a period of reorganisation in which many UK firms will realise that they cannot maintain reasonable levels of profitability in a deregulated world. Many smaller stockbroking firms will rationalise or be forced to merge, as was the case in New York in 1975, unless they have formulated a

*Euromoney Publications, London 1984.

carefully planned strategy. Niche houses will prosper. Some of the UK merchant banks, who have elected not to provide the full range of financial services after deregulation, will pursue a specialist style similar to Lazard Frères in New York. The US and Japanese behemoths are not especially interested in the UK retail market. Those firms who have an established private client base and an effective control of cost overheads, will not be unduly threatened by the overseas predators. UK research expertise, which is probably second only to New York, should be broadened to include selected European shares, which today attract only superficial domestic analysis. The UK domestic debenture market and the Unlisted Securities Market (USM) are hardly the intended preserve of the international newcomers. Contrary to popular opinion there will not be blood flowing on the streets from London's former financial élite.

The determining factor for survival and success will be in the provision of service. Flexibility will also be essential. London has the advantage of its obvious familiarity with its own home market. The disadvantage is that many of London's financial institutions have fallen far behind in new product innovation, which has become almost a trademark of their overseas competitors.

Does London have a coherent long-term strategy for the future? The Financial Services Bill is well intentioned, but its objectives place far too much emphasis on the protection of the retail investor rather than guaranteeing that London, together with Tokyo and New York, will be one of the three financial centres of the world. Fortunately, London will be able to accomplish this without undue difficulty if unnecessary bureaucratic interference can be kept to a minimum. In a speech early in 1986, Ian Steers, Chairman of Wood Gundy & Co in London, reminded his audience of London's natural advantages:

1 the availability of skilled personnel;
2 language compatibility;
3 a high ethical standard in the market;
4 facilities (office space, good transportation, etc.);
5 an advantageous time zone;
6 a traditional background of international trading in other markets such as shipping, banking, insurance and commodities; and
7 a positive and welcoming government attitude.

He continued by saying, 'The infra-structure which is in place is so big and the number of people directly involved so large, that only a major change in government policy as to tax or regulation could cause the market to move, fragmentation as we had it in the 1960s and early 1970s seems unlikely and, in my view, only one centre could provide similar facilities, that is New York, where the time zone is unfavourable but otherwise with some happy intervention by the authorities could provide an alternative base, which is not to say that our friends in Amsterdam, Zurich and Paris are not taking steps to make themselves more attractive to international trading of securities.'

In other words, London must look to its laurels in order to ensure its place as one of the three main financial centres of the world. Ian Steers' last comments on London's natural advantages of 'a positive and welcoming government attitude' would not be accepted by everyone. First, the City has only just admitted foreign membership of the London Stock Exchange and this in itself was the result of a *fait accompli* rather than the laying down of a welcome-mat. Secondly, the initial proposals for the self regulatory organisations were so clumsy and their accountability to the new Securities and Investments Board so vague, that the majority of foreign financial institutions in London despaired. While these have since been refined and improved, there are many who consider that some of the remaining regulations are negative rather than positive.

THE CITY AND ITS HISTORY

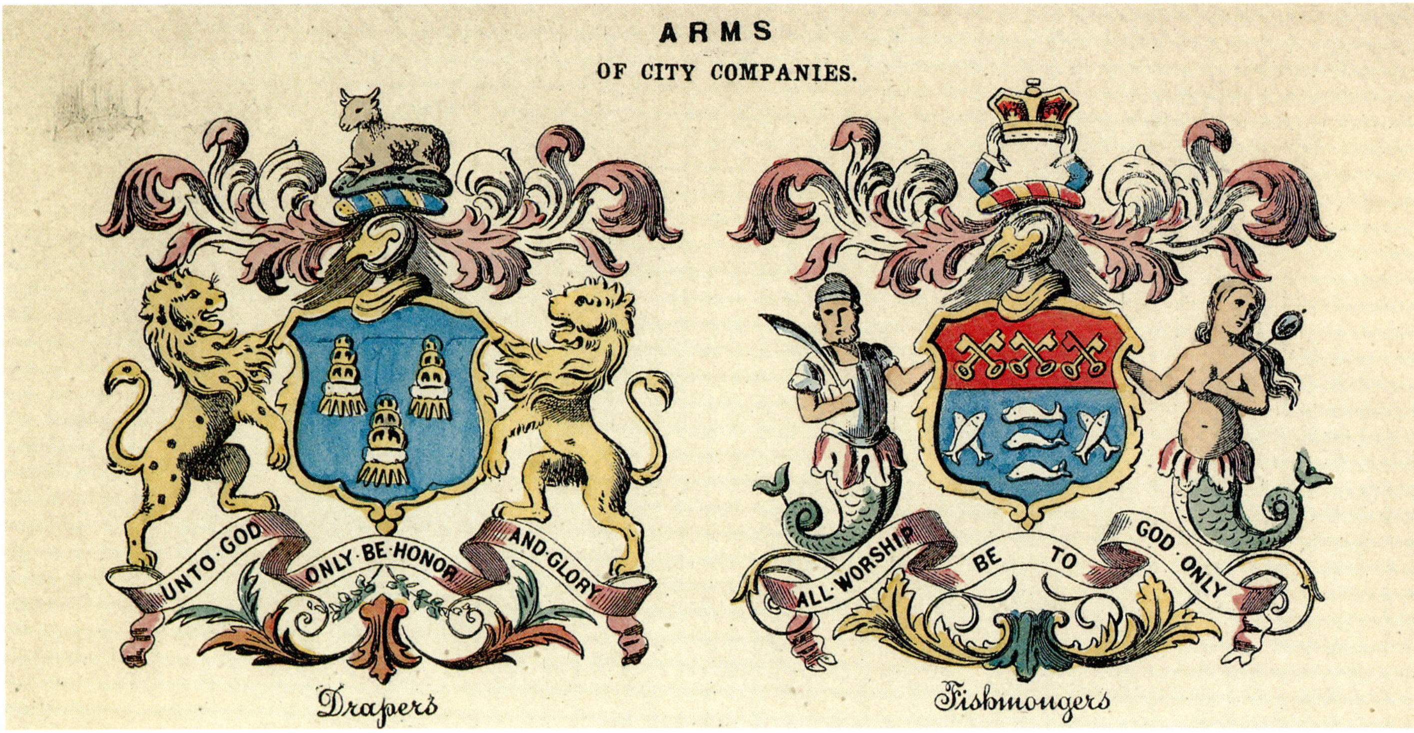

Above, and pages 22-23: the arms of the 12 great livery companies

CHAPTER ONE

The City – a background

The outsider's perception of the City of London is still one of mystery. It is not a 'City' in its own right. It does not promote any specific public image. To be 'someone in the City' is a legacy inherited from countless mothers and aunts who are at a total loss to explain the exact career of their family. The only common denominator of the modern 'City' is finance. To some extent this is true. Almost every function of the City revolves around money. For all its good intentions and charitable works the fact is that the City of London is little more than a money machine. Money is there to be raised, made and lost. The Armorial Bearings of the City of London, first recorded in April 1381, carry the motto 'Domine Dirige Nos'. A more contemporary view of the City might include the motto 'In pursuit of profit'.

The City's forefathers would not like this interpretation. They would point, with justifiable pride, to the history and tradition of the City. This was the Roman's Londinium, a site chosen not for the beauty of its natural habitat but more practically for its limitless supply of fresh water. Was Londinium not, according to the description of Tacitus, 'a town of the highest repute and a busy emporium for trade and traders'? In earlier days it was also a place to live. Before the Great Fire of London in 1666 a quarter of a million people lived in the area which now incorporates the City. The subsequent move west did not significantly change its influence. Despite the fact that no more than 6,000 people live in the City today, it retains a remarkable degree of autonomy. It has its own boundaries, police and courts, as well as its own Lord Mayor, aldermen and councillors. During working hours the City's population rises to more than 500,000.

Why do almost half a million people flock to the City every day? The answer is that the City money machine requires an enormous array of individual skills to keep it in motion. Not everyone in the City is in pursuit of profit, but it is no coincidence that the City of London is almost the only area in the United Kingdom which can boast of minimal unemployment.

A London-based merchant or investment bank announcing a takeover bid on behalf of a client will draw on many of the services which the City can provide as a matter of course. The merchant bank initiates the takeover. Its public relations firm begins to woo the press and the shareholders of the target company. Its stockbrokers buy shares in the open market. The lawyers begin drafting the takeover documents. Financial analysts begin examining the takeover terms to see if they represent good value. The printers are informed of the total documentation that will be required. The financial press contact both the takeover target and the prospective acquiror to listen to their immediate strategy. The investment professionals begin to take huge long or short positions estimating the outcome of the takeover. The Takeover Panel and the Stock Exchange will look at price behaviour patterns to establish whether insider trading could have taken place. The Takeover Panel and the Monopolies Commission will start to look at the strategies being employed and the implications of a successful merger within specific sectors of industry. A chain of events has been set in motion which immediately involves thousands of people from a wide variety of professions.

In the City it is rarely possible to stray far from the central theme of finance. True, there are theatres, the Guildhall School of Music and Drama and an Arts and Conference Centre. However, no-one is under any illusion as to who pays the bill. The City generates annual earnings of about £2,000 million per year. Its rents and rates are among the highest in the country. The City's own independent wealth is used to endow projects far beyond its own boundaries.

The ancient City ceremonies which have been conducted for centuries are still observed, but are really little more than a pleasant diversion from the main pre-occupation of finance. Most of the City's younger generation of workers are aware of the tradition of the livery companies, if only because so many streets and halls are named after them. However, few would probably be able to name the 12 'great' livery companies: the Mercers, Grocers, Drapers, Fishmongers, Goldsmiths, Skinners, Salters, Haberdashers, Merchant Taylors, Clothworkers, Ironmongers and Vintners. The traditional City ceremonies of the livery companies also mean little to the new generation of dedicated professionals motivated by personal recommendation and by the prospects of promotion to senior management positions. The tradition of the Worshipful Company of Butchers presenting the Lord Mayor with a boar's head may seem almost trivial to many City practitioners. Nevertheless, this is an example of the City's history at its classical best because it represents a 640-year-old transactional agreement based on the simple consideration of commerce. In 1343 the City gave a plot of land on the Fleet River to the butchers of St Nicholas Shambles for the purpose of 'cleansing the entrails of beasts'. In return they were annually to present the Lord Mayor with a boar's head. Even when the butchers moved to new premises the tradition continued.

Similarly, ceremonies such as the Trial of Pyx, Swan Upping and

Road Sweeping by wine porters of the Vintners Company are still conducted to remind the City of its heritage.

The emergence of the City

Beyond the pageantry, however, the connection with business and commerce was always evident. The Stock Exchange began to evolve in the 1670s through the steadily rising number of transactions conducted in the region of the Royal Exchange. Lloyds of London, still the world's premier insurance market, had its origins in Edward Lloyds Coffee House in 1687. Perhaps the most important issue of the time was the change in political philosophy of the seventeenth-century nations. The importance of commerce was only just being recognised. Territorial advantage was of little use on its own unless accompanied by trade advantage. The central basis for previous wars had been religion. In the new era wars would be fought for trade expansion. It was therefore only a natural progression that the City of London with its natural ties to commerce should be asked to finance the wars that would provide England with new territories which would in turn offer new trading opportunities. The Bank of England was incorporated by Act of Parliament and Charter in 1694. In return for receiving the exclusive right to issue bank-notes in England and Wales it provided funds for William III's war with the French.

As the British commercial empire expanded through the acquisition of new overseas territories in the eighteenth century, the role of the City of London became even more important. As a financial centre London was now almost the equal of Amsterdam, which could boast the first Stock Exchange and the first central bank, but London's ports handled more cargo than the Dutch capital. Within a relatively short time the Spanish–Dutch wars of independence would diminish Amsterdam's importance to a fraction of its former level. In those days of mushrooming trade and the huge wealth which it created there was inevitably some speculative excess. London's early stock markets were often the subject of price manipulation and outright fraud.

The South Sea Bubble

By far the most famous scandal was the South Sea Bubble, but it was the speculative fever of the early eighteenth century which created the classic environment for a period of financial folly which may never be surpassed. In his brilliant book *Extraordinary Popular Delusions and the Madness of Crowds*, which remains a best-seller today, the Victorian author Charles Mackay recounted some of the excesses of that particular period after the South Sea Company was founded by the Earl of Oxford in 1711. Some historians may have forgotten that the original objectives of the South Sea Company were considered to be entirely commendable, and contemporaries spoke of the 'Earl of Oxford's masterpiece'. The South Sea Company effectively took over part of the national debt, amounting to around £10 million. The government agreed to pay annual interest of 6 per cent secured by the duties on a wide range of goods including wines, vinegar, Indian products, wrought silks, tobacco and whale fins. However, the speculative attraction of the venture was that the company was granted a monopoly of trade to the South Seas.

The gold and silver mines found in Mexico and Peru were already legendary and the general public believed that they were inexhaustible. The South Sea Company held out the prospect of being able to open Spanish-owned ports to English traders who would reap huge profits. In fact, Spain never had any intention of allowing free trade in her possessions and the volume of subsequent commercial transactions was minimal. Nevertheless, the company's reputation remained high and in 1720 the South Sea Company offered to underwrite the entire national debt of more than £30 million despite the unreserved criticism of Robert Walpole who recommended that this role should be implemented by the Bank of England. The price of the South Sea Company stock rose from 130 to 400 before settling back at around 350.

In the House of Lords many prominent peers spoke against the scheme, including Earl Cowper, who compared the new proposed legislation in favour of the South Sea Company to the horse of the Siege of Troy, which was welcomed by everyone but contained treachery and deceit. When the legislation received the Royal Assent, the roads around Exchange Alley and Cornhill became impassable as crowds gathered to buy South Sea Company stock.

It was only after the Royal Assent had been given that artificial manipulation of the South Sea stock began. Initially the price fell, but as rumours, usually without any foundation, were circulated that the English might even cede Gibraltar and Port Mahon for ports and unlimited trading rights in Peru, the rise was resumed. The South

The South Sea Company offered to underwrite the entire national debt of more than £30 million

Sea Company announced that it would increase its capital by a further £1 million which was immediately oversubscribed. The South Sea Company's public subscription tranche was payable in five parts, an innovation that was only to be repeated in the Eurodollar bond market in the form of partly-paid bond, 260 years later, in the summer of 1980. The inherent leverage in partly-paid bonds was certainly in evidence in 1720, where the value of the initial partly-paid stock doubled in value within a few days. Demand was such that the South Sea Company offered a further £1 million to the public which was, again, immediately oversubscribed.

In a time of such feverish speculation there were inevitably many imitations. Some of the ventures were serious, but others were designed purely for the gullible. Maitland's *History of London* reminds us that one of the most encouraging projects was the establishment of a company to make deal boards out of sawdust. Mackay recalls a company which would design 'a wheel for perpetual motion' and, perhaps the ultimate in absurdity, 'a company for carrying on an undertaking of great advantage but nobody to know what it is'. The fact that even this issue was oversubscribed only underlines the prevailing mood of the time. To give the promoter of this company his due, he promised investors wealth beyond the imagination of almost anyone in eighteenth-century England. At at time when annual rates of interest rarely exceeded 5 per cent, the company operating in areas of 'nobody to know what it is' offered returns of 100 per cent and this for an outlay of only 2 per cent in partly-paid form. The shares were in denominations of £100 each but the initial down-payment was only £2 per share. Every subscriber who paid his deposit would then be entitled to a dividend of £100 p.a. when the remaining £98 for the fully-paid shares had been received. Of course, it was no more than the equivalent of a modern day scam. One-thousand subscribers had paid £2 per share down-payment, and with this not insubstantial amount of money, probably worth a minimum of £500,000 equivalent today, the promoter departed for the Continent on the following morning.

There were so many 'Bubble' companies and ventures that it would take too long to list them all but the selection shown below will indicate the level of speculation which had extended to all levels of society in England:

1 For building and rebuilding houses throughout all England. Capital, three millions.
2 For effectively settling the island of Blanco and Sal Tartagus.
3 For improvement of Flanders lace.
4 For encouraging the breed of horses in England, and improving of glebe and church lands, and for repairing and rebuilding parsonage and vicarage houses.
5 For improving the land in the county of Flint. Capital, one million
6 For purchasing lands to build on. Capital, two millions.
7 For trading in hair.
8 For carrying on an undertaking of great advantage; but nobody to know what it is.
9 For paving the streets of London. Capital, two millions.
10 For furnishing funerals to any part of Great Britain.
11 For erecting loan-offices for the assistance and encouragement of the industrious. Capital, two millions.
12 For insuring of horses. Capital, two millions.
13 For a grand dispensary. Capital, three millions.
14 For improving the art of making soap.
15 For a wheel for perpetual motion. Capital, one million.
16 For insuring and increasing children's fortunes.
17 For trading in and improving certain commodities of the produce of this kingdom, &c. Capital, three millions.
18 For paying pensions to widows and others, at a small discount. Capital, two millions.
19 For a grand American fishery.
20 The Bottomry Company.
21 For drying malt by hot air.
22 For employing poor artificers, and furnishing merchants and others with watches.
23 Another for the improvement of our breed in horses.
24 Another for horse insurance.
25 For erecting houses or hospitals for taking in and maintaining illegitimate children. Capital, two millions.
26 For making iron with pit coal.
27 For buying and fitting out ships to suppress pirates.
28 For the transmutation of quick-silver into a malleable fine metal.

There are many other enjoyable stories of this time, including the 'Puckles Machine Company' for discharging round and square cannon-balls and bullets. This company's promoters assured would-be subscribers that it would revolutionise the art of war, but

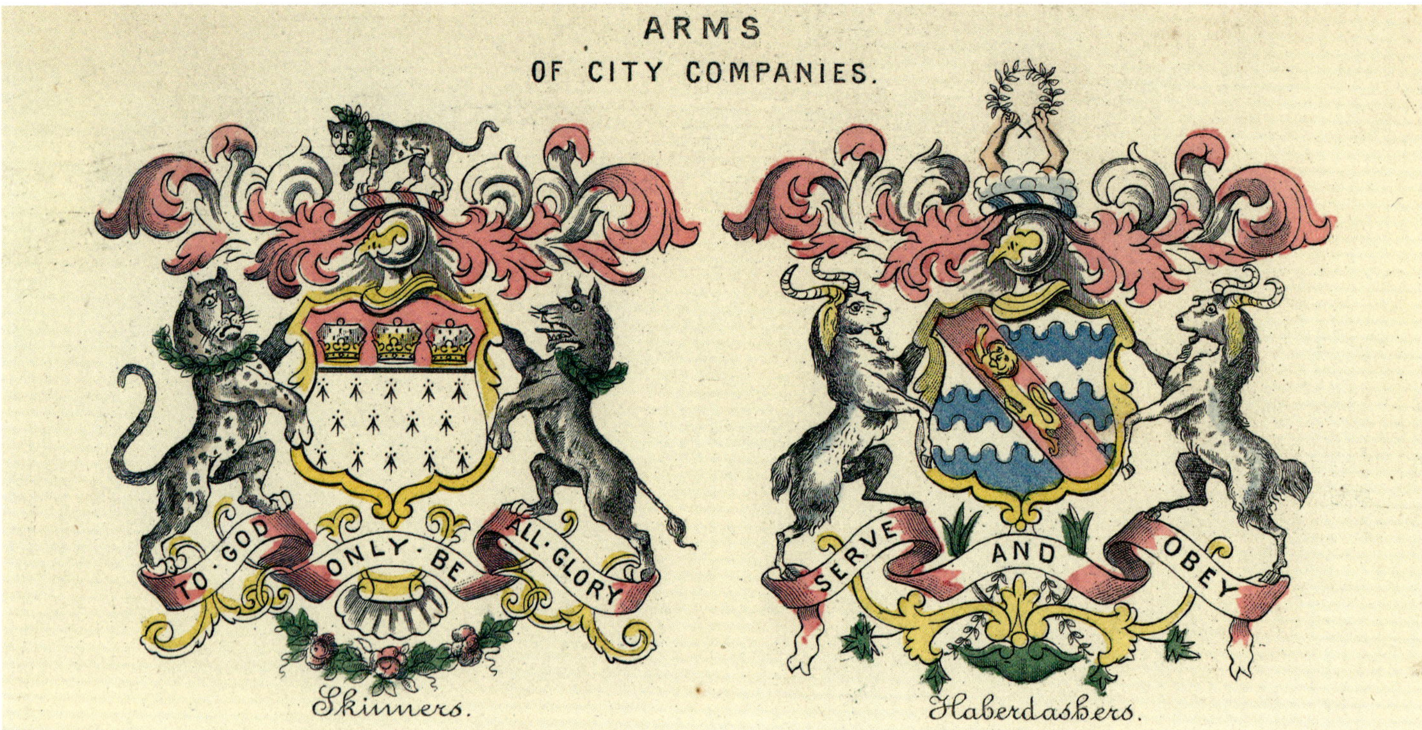
ARMS
OF CITY COMPANIES.
TO·GOD ONLY·BE ALL·GLORY
Skinners.
SERVE AND OBEY
Haberdashers.

SAL SAPIT OMNIA
Salters.
GOD·IS OUR STRENCTH
Ironmongers.

Goldsmiths

Merchant Taylors

Vintners.

Clothworkers.

the cartoons and the news-sheets of this era show that the speculation frenzy was openly recognised. The 'Puckles Machine Company' was therefore described as:

> A rare invention to destroy the crowd
> Of fools at home instead of fools abroad.
> Fear not, my friends, this terrible machine
> They're only wounded who have shares therein.

The bubble could, however, only burst with the fortunes of the South Sea Company which had initiated the speculation. Between 1 May 1720 and 3 June 1720, the price of the South Sea Company stock rose from 500 to 890. In August it reached 1,000 per cent, but had fallen back to 700 a month later when it became known that the Chairman, Sir John Blunt, and other directors had sold their personal holdings. Despite a public meeting held in the Merchant Taylor's Hall in September, at which the directors professed total confidence in the company, the price of South Sea Company stock fell to 640 on the same day and then continued consistently downwards to 135. Many bankers and goldsmiths who had lent money against South Sea Company stock were ruined, and the company's principal cashiers suspended payments.

The bubble had lasted less than a year. Parliamentary history records:

> And thus were seen in the space of eight months the rise, progress and fall of that mighty fabric which being wound up by mysterious springs to a wonderful height had fixed the eyes and expectations of all Europe, but whose foundation being fraud, illusion, credulity and infatuation fell to the ground as soon as the artful management of its directors was discovered.

The recriminations against the directors were to some extent justified, but they did not take into account that the nation itself had been equally guilty of avarice. Although the company's secretary had fled the country with numerous records, it was soon revealed that the directors had given themselves and others preferential subscriptions of stock. Very few of the directors, however, were committed to prison and within a few years confidence in public credit was restored. Speculation on a similar scale did not take place until more than a century later with the railway construction boom in the mid-1840s.

For all the furore at the time of the South Sea Bubble, that particular period of speculation does not seem to have had any

Hogarth's satire of the South Sea Bubble

What is remarkable was that the great leaders of the financial world took no part in the railway boom

lasting effect on the City's reputation. Investors became more cautious. The demand for government loans became more pronounced. In 1781 a number of separate 3 per cent annuities were integrated into a Consolidated Bank Annuities Fund, thereby creating the famous 'Consols'. Two undated Consols issues are still traded. With England almost constantly at war during this period, and with the information systems painfully slow, prices would fluctuate substantially on almost any rumour. In 1797 the price of the 3 per cent Consols fell from 97 per cent to under 50 per cent upon rumours, which proved to be totally unfounded, that the French army had landed in Wales. Mayer Rothschild's stockmarket coup, in the aftermath of the Battle of Waterloo in 1815, has been doubted by some historians, but what is not disputed is that the house of Rothschild had a better information system than almost any of its competitors.

The Industrial Revolution

After the Napoleonic wars the City of London enjoyed a period of prolonged prosperity, assisted by the coming of the railways and the huge fortunes made during the Industrial Revolution. London was recognised as the financial centre of the world. Because sterling, which was exchangeable into gold, was regarded as the strongest currency in the world, London attracted international capital seeking a secure investment. London also provided the world's largest market for borrowers seeking long-term capital. The goods manufactured in the Industrial Revolution had a ready market-place in Britain's colonial possessions.

A number of historians have suggested that the City of London and the banks provided only limited finance for the Industrial Revolution. That the merchant banks 'provided funding for the building of every railway in the world apart from their own' was a popular criticism. It is true that the great family merchant banks were international in their outlook, because the majority of the founders came from Continental Europe. The Hambros, Lazards, Rothschilds and Schroders had all come to London from Europe, and in several instances other members of the family opened banks in alternative financial centres, such as London and New York. These banks were very experienced in the accepting and discounting of bills on overseas customers. Domestic lending was largely the reserve of regional banks. In terms of influence and power, the merchant banks probably reached a peak during the Napoleonic wars, when they were responsible for funding a significant portion of the cost.

With the cessation of the wars with France, the natural gravitation of the banks was to third-country financing to support imbalances of trade. The City of London became the major supplier of capital to the world. In 1822 six Stock Exchange members requested the Committee that a market in foreign securities be created which would afford 'great convenience and profit to the House at large'. The British merchant banking houses might have been better advised to finance the principal domestic railway systems, many of which yielded very large profits to the original subscribers. Almost inevitably the railway boom resulted in its own period of speculative excess. The spate of regional railway company failures, including several which had not even started initial construction, led to the Companies Act of 1844, which required that all new joint-stock companies must be registered. Provincial bank failures exacerbated the problems of the time. Disraeli himself pointed out the lack of City support for the railway boom, as did its chief architect George Hudson, when he was quoted as saying 'what is remarkable in this vast monument in which so many millions were produced and so many more promised was that the great leaders of the financial world took no part in it'.

The financial lists of the time tell their own story. The majority of new listings were for foreign borrowers or entities. Railway obligations, in countries throughout the world, were a major component of all securities trading. They were not always to prove a sound investment. The great house of Baring Brothers had to receive temporary financial assistance from the Bank of England in 1890 due to problem loans in South America. In the United States many railway companies offered to the public were blatantly fraudulent. However, UK capital flows abroad continued to increase.

The turn of the century saw some spectacular bankruptcies in the United States, but this was also the time of the grandiose industrial trust companies which were, in effect, in a monopoly position in several major industries. In the winter of 1901 John Pierpont Morgan engineered the amalgamation of Federal Steel, Carnegie Frick, National Tube and various wire companies. The result was the first US$1 billion coporation, US Steel Co. The successful amalgamation required the consent of steel magnate Andrew Carnegie. The following extract from Andrew Sinclair's book

THE BULLS AND BEARS ALARMED.

Corsair on the life of John Pierpont Morgan shows the comparative size of the transaction in 1901 terms.

> It is hard to imagine what a billion dollars meant at the time. It represented a twenty-fifth of the whole national wealth, more than the combined dividends from the railroads for eight years, and more than the value of all the wheat and barley and cheese and gold and silver and coal produced in 1900 in America. It would have built 20 Krupp steel plants, the largest in Europe and the pride of Germany. No merger on this gigantic scale had yet been attempted. The question was not would it work, but would Carnegie sell out. Schwab was doubtful because he knew the changeable moods of the old Scotsman of 66. 'Well', Morgan said in the dawn light, closing the conference, 'if Andy wants to sell, I'll buy. Go and find his price.'

Morgan was successful. He bought the Carnegie interests for $420 million, of which Andrew Carnegie received $213 million to add to his fortune already estimated at $100 million. By any standards it is an astounding sum of money when translated into present-day terms. On concluding the transaction Morgan shook Carnegie's hand and said, 'I want to congratulate you on becoming the richest man in the world.' Lord Rothschild remarked that Carnegie had amassed a fortune in one generation equivalent to that earned by the Rothschild family in a century. Later Carnegie confided to Morgan that he could have asked for another $100 million. Morgan immediately agreed.

With such exciting prospects abroad, it was no surprise that UK investment capital flowed abroad. Although the investments were more speculative, better returns could be achieved overseas. London's position as the centre of the financial world was reinforced by the rise of Lloyds of London, which became the dominant force in insurance. Lloyds received statutory backing in 1871 through the Lloyds Act.

The two World Wars

The strategic role of sterling and its pre-eminent position in the financing of world trade had almost no serious challengers until the First World War. Because of its convertibility, sterling was 'as good as gold'. At the outbreak of war the Accepting Houses Committee was formed initially to provide financial assistance to houses with assets frozen or debits due in now hostile countries.

The period between the First and the Second World War took in the stock market crash of 1929 in the United States, the Great Depression of the 1930s and the, in retrospect, totally wrong decision of the United Kingdom to return to a gold standard between 1925 and 1931. Surprisingly, the City was able to prosper and expand during this most difficult period. Because sterling, due to its non-convertibilty, was no longer as good as gold, the City developed its own expertise in foreign currency trading, which later was officially formed into the Foreign Exchange and Currency Deposit Brokers Association. During the Second World War, foreign exchange rates were fixed, which left sterling, despite the Allied victory, as a significantly overvalued currency after the war.

With sterling clearly overpriced, and the government immediately imposing foreign exchange controls, the outlook for the City

New York's crash also brought out the crowds in Throgmorton Street, 23 September 1929

Early investments made in Japan have, in some instances, risen in value by more than 100 times

appeared bleak. The Forex market itself was not re-opened until 1951, and even then dealings were restricted. Again the City showed its resilience in times of adversity. The insurance industry was a constant source of high invisible earnings. British fund managers continued to invest substantially abroad, despite the penal investment currency premium. Early investments made in Japan when the stock markets were re-opened after the Second World War have, in some instances, risen in value by more than 100 times. Niche sectors such as arbitrage in South African gold shares proved to be extremely profitable for the specialist international houses. Traditional barriers were being broken down in the City's establishment. Hostile takeovers were not totally new, but in the 1950s they gained increasing recognition as well as widespread criticism.

In 1953 Harold Samuel had attempted to acquire the Savoy Hotel group, an ambition which several others had failed to achieve more than 30 years later. Samuel bought 20 per cent and the late Sir Charles Clore 9 per cent, but the bid was resisted with some ease due to the complex voting structure of the Savoy group. The acrimony which ensued forced the Bank of England to again request banks and financial institutions not to finance unwelcome takeovers. In 1959 Clore made his highly controversial bid for the Watney brewing group, comprising five breweries and 3,670 public houses. That bid was also a failure, but it made the City aware that huge unrealised surpluses were contained in the balance sheets of companies with substantial property assets. This, in turn, was sufficient to inspire the property boom in the early 1960s, which was to be repeated a decade later, causing a major banking crisis.

The Euromarkets

The City's most important achievement in the late 1950s and the early 1960s was to develop a non-sterling market. The new 'Euromarkets', which were then so unfortunately ignored by the City, provided the almost identical base upon which the new deregulated London markets are now being built. 'Euro' has never meant 'European', but rather a true cross-boundary international market. So many people have claimed to have invented the Euromarkets, that it is worthwhile to record the comments of one of the first acknowledged Euro specialists, Paul Einzig, in his book *The Euro-Dollar System.*

> Through constant usage in the small but expanding literature on the new system, the term 'Eurodollar' has come to convey much more than mere indication of geographical location. It has come to convey something even more than just deposits in terms of a foreign currency. Such deposits had existed since time immemorial, but Eurodollars are something different. A Eurodollar market, in the sense in which the term is understood, has only been in existence since 1957, but there have been innumerable precedents for accepting deposits in terms of some foreign currency. Those who seek to minimise the new development are of course in a position to point out that already in the early 1950s Italian banks were borrowing systematically dollar deposits from French banks, and that about the same time, or possibly even earlier, Communist-controlled banks adopted the habit of camouflaging their ownership of some dollar deposits by redepositing them with European banks. Bankers whose memories go back to the inter-war period may remember that during the 1920s there was active business in sterling and dollar deposits in Berlin and Vienna.
>
> Some students of banking history succeeded even in tracing back transactions in sterling deposits in foreign countries to before 1914, producing concrete evidence about their existence in the late nineteenth century. In my own *History of Foreign Exchange* I recorded the medieval practise of drawing bills payable at the quarterly fairs in terms of foreign currencies other than those of the countries of payment. It seems reasonable to assume that, since business was actively transacted in foreign currencies at such fairs, such bills were often paid out of or into foreign currency deposits. A thorough historical research would probably disclose many more definite early precedents. Even so, the conclusion inferred from such precedents, that the market in Eurodollars with which we became familiar in the late 1950s was not an innovation, seems to be entirely unjustified.

Most historians now credit the late Sir George Bolton, Chairman of the former Bank of London and South America, for developing the Eurodollar deposit market, and the late Sir Siegmund Warburg, Chairman of S. G. Warburg & Co, for developing the public Eurodollar bond. The argument is essentially academic when viewed against the enormous benefits which these markets provided for the City. Foreign banks flocked to London to tap the Eurodollar pool. It was the Eurodollar pool which indirectly made London the centre of the syndicated loan market which, until the late 1970s, was the largest single source of major financing for sovereign credits.

This was the era of asset-stripping in its most unacceptable form

The syndicated loan market subsequently declined in importance and its position has been overtaken by floating rate notes, note facilities and commercial paper programmes. The Eurobond markets have gone from strength to strength and, with total primary market volume of US$133 billion equivalent in 1985 and probably US$160 billion equivalent in 1986, they now dwarf the London markets as a capital-raising mechanism.

While the embryo Euromarkets were flourishing, the London Stock Exchange found its growth inhibited by being a single-currency market whose domestic investors were precluded, unless they were willing to pay a premium, from buying foreign currency securities. The market's volume was therefore not supported to any extent by foreign investor interest. Overseas buyers believed that sterling was vulnerable, a view that was confirmed by the November 1967 devaluation of sterling under Harold Wilson's Labour government.

Property boom and takeovers

The return to power of the Conservative Party caused a major revival in the stock market, assisted by another property boom and a surge in the number of take-over bids. This was the era of asset-stripping in its most unacceptable form. The property boom had been fuelled by Chancellor Anthony Barber's cheap money policy. When the tide turned and the asset-strippers found themselves faced with rising interest rates and falling property values, the result for many was ruin. By 1973 stock markets were already in a decline, but in the City the decline became a crisis. The 1973 Arab–Israeli war triggered the beginning of an upwards spiral in oil prices that was to continue until the early 1980s. With energy-related products comprising the majority of consumer price indices, inflation soared. In the United Kingdon a large group of secondary or 'fringe' banks were seriously overlent on speculative property development. The Conservative government under Edward Heath took a stand against the wage demands of the National Union of Mineworkers, but the resulting strike brought the country almost to a standstill.

In February 1974 Labour was returned to power, albeit by a small majority which depended upon the support of a few Liberal Members of Parliament. The Labour Party blamed the City for rising unemployment due to the City's apparent inability to provide permanent capital for investment. The accusations did affect confidence in the Stock Exchange, where share prices were starting to collapse. On 16 August 1972 the Financial Times Ordinary Share Index had peaked at 537.9, and by 6 January 1975 it had fallen to a low of 146.8. If confidence in the City was not high, there was also considerible despondency within the City itself. Between July 1974 and August 1975 nearly 70 Stock Exchange

Sir Siegmund Warburg developed the public Eurodollar bond market

member firms had either merged or ceased trading.

It would perhaps be unfair to suggest that the Stock Exchange was slow to improve itself in the era when oil prices quadrupled over a very short time, and international markets everywhere were in deep depression. Stock market prices did recover sharply in the 1976–77 period and falling interest rates helped a revival in international capital markets. Sterling, was, however, still considered to be a weak currency, and another crisis duly occurred in September 1976 when the pound fell to US$1.76. With sterling in retreat and a Labour government in power there was little foreign investment interest in the City, and very little sympathy from Whitehall.

Restrictive trade practices

The Stock Exchange itself first came under attack with the passage of the Restrictive Trade Practices Order of 1976, itself a successor to a 1973 Act drawn up by a Conservative government with a view to examining existing cartels and encouraging competition. While the legislation was not intended to include self-regulatory bodies, the Stock Exchange found that it was not one of the bodies deliberately excluded. The intentions of the Act, which was passed under a Labour administration with little affection for the City, were to make all restrictive trade practices illegal. Among the specific areas defined in the Act was the prevention of competition. When the Stock Exchange registered its Rules and Regulations and its Dealing Code with the Office of Fair Trading, it was subsequently informed that a number of restrictive practices had been identified.

Perhaps the most important fact was the alleged restrictive practice which limited the freedom of brokers to supply full broking services to their clients. Inevitably, this again resulted in a debate on the question of fixed minimum commissions. Another alleged restrictive practice related to the basic market system of operation, with the division of jobbers and brokers into single-capacity functions. When the Director of Fair Trading referred the supposed restrictive practices to the Restrictive Practices Court, the Stock Exchange had no alternative but to take legal advice.

From the outset there were doubts whether the Restrictive Practices Court was the ideal platform to judge whether the Stock Exchange system operated against the public interest. The conclusions of the Wilson Committee, formed in 1977 under the chairmanship of former Prime Minister Harold Wilson, doubted the suitability of the Restrictive Practices Court. So, too, did the Bank of England, which lobbied hard but unsuccessfully that another forum would be more appropriate. The Chairman of the Stock Exchange, Sir Nicholas Goodison, did not criticise the basic investigation but preferred to see the appointment of a Royal Commission, which would make recommendations for greater efficiency. In order to extricate itself from the court proceedings, the Stock Exchange was advised by its legal counsel that this could be achieved through intervention by the Secretary of State for the Department of Trade and Industry. It was, however, essential that the intervention took place before the Case for Prosecution through the Restrictive Trade Practices Court had been completed.

The return of a Conservative government in 1979 gave the Stock Exchange's supporters some reason for optimism. Mrs Thatcher's first Trade Secretary, John (now Sir John) Nott, the current Chairman of Lazard Brothers and a former banker with S. G. Warburg & Co, was sympathetic, but he was aware that any quick assistance for the Stock Exchange would be interpreted as dispensing favours to the Conservative Party's friends in the City. Mrs Thatcher, too, was sensitive of the new government's image.

Sir John Nott, as Trade Secretary in 1979, was sympathetic towards the Stock Exchange but sensitive of the government's image

In October 1979 court proceedings were served on the Stock Exchange. The progression of events began with the Stock Exchange preparing its opening Statement of Case, which was finally delivered to the Court in March 1981. The Office of Fair Trading then produced its answer to the Statement of Case, and the Stock Exchange had then to make a further plea. The complexity of the case was evident in Sir Nicholas Goodison's letter in 1981 to the Stock Exchange members, which stated that he did not anticipate a response to the Statement of Case until the following year.

Throughout this period the Stock Exchange and the Bank of England lobbied their case with successive Secretaries of State for the Department of Trade and Industry. The Bank of England found a more responsive attitude in Lord Cockfield, who was subsequently succeeded by Cecil Parkinson after the landslide Conservative victory in mid-1983. Parkinson proved to be more approachable on the Stock Exchange's troubles. An accountant by training, he had been the architect of the Conservative victory and was regarded as one of the best intellects in the party. More importantly, the new Trade Secretary had foreseen the need for change within the City. 'The process of City reform should have started in 1979 after the Conservative Party removed foreign exchange controls,' he said. The Bank of England's argument to remove the case against the Stock Exchange from the Restrictive Practices Court was reinforced by an internal study undertaken by the Bank itself. This showed that the overall performance of the City in relation to its principal foreign competitors was less than impressive. Overseas financial institutions dominated the Euromarkets. The respectable, rather than outstanding, share of the international syndicated loan market held by British banks had to be viewed against the fact that this market was in the process of a major decline, due to the introduction of revolving underwriting facilities, note insurance facilities, commercial paper programmes and other short-term money market instruments. In insurance Lloyds of London was still the centre of the industry, but recent scandals had brought Lloyds into serious disrepute. However, the real problem was the Stock Exchange. Although the United Kingdom is arguably now the world's major centre for international fund management, the City had failed to make a significant impact in the international capital markets. In essence, there appeared to be some substance behind the accusations that the London Stock Exchange had been standing still.

The study by the Bank of England may have supported the Bank's lobbying effort, but it was also disturbing. If the London Stock Exchange had been unable to perform well at a time when it effectively operated under a cartel, what would be the consequences of reforms which included allowing foreign financial institutions entry to the market-place? The Euromarket's example was distinctly ominous. However, if there were no reforms, the City was in danger of running a very poor third to New York and Tokyo as a world financial centre. The Bank of England had strong reservations, which it still holds, concerning the introductions of reforms which would almost certainly result in the early US takeover of large parts of lucrative securities business, which had formerly been the preserve of London member firms. It was all very well for the members of the Stock Exchange Council to protest publicly that foreign financial institutions would not quickly dominate their business. This was bad public relations because, within a year, almost all the Stock Exchange Council members had seen their firms bought out by financial institutions, including a large number of foreign houses. For the Bank of England it must have seemed like a case of 'Morton's Fork', through which Henry VII's Cardinal Morton levied taxes on the British aristocracy, on the basis that if

Cecil Parkinson as Trade Secretary asked the Stock Exchange in July 1983 to draw up proposals for reform

If they displayed wealth they could afford taxes and if they appeared impoverished they were clearly hiding their wealth

they displayed wealth they could afford taxes and if they appeared impoverished they were clearly hiding their wealth. The Bank knew that even if the Stock Exchange won its case in the Restrictive Practices Court, it might then retreat back into its own cocoon of insularity and undermine the City's role as a financial centre. If changes were introduced, the City would be exposed to the full force of foreign competition for which it was inadequately prepared.

Proposals for reform

In July 1983 Cecil Parkinson asked the Stock Exchange to draw up proposals for reform which would enable the Conservative government to move the case against them from the Restrictive Practices Court. That the Stock Exchange was able to respond with four basic proposals by 22 July confirmed it had already accepted that change was inevitable. The proposals were:

1 To dismantle the rules prescribing minimum scales of commission by 31 December 1986 at the latest.

2 To establish an independent Appeal Tribunal which could overrule the Council's decision to reject an applicant for membership.

3 To introduce lay members to the Council and to introduce to the present Appeal Committee a majority of individuals who were not Council members.

4 To introduce rules which permitted non-members to serve as directors of corporate member firms, providing that the majority of directors were members.

On the same day Cecil Parkinson announced his acceptance of the proposals. He pointed out in the House of Commons that the court proceedings under the Restrictive Trade Practices Act may not have been the best method of pursuing the points raised by the Office of Fair Trading, under its Director General Sir Gordon Borrie. The Stock Exchange was unable to press ahead with changes which would allow its members to compete for international business, and there was a danger that the legal proceedings would affect the efficient operation of the Stock Exchange itself. The government would therefore wish to see the matter settled out of court. In addition, the government had concluded that investors would be best safeguarded by maintaining the separation of the function of jobbers and brokers for the time being. A reference was made to the fact that the House had already agreed to separate the role of insurance underwriters from insurance brokers.

Criticism of the decision to remove the case from the Restrictive Practices Court was inevitable, particularly from the Labour Party opposition who had essentially engineered the legal proceedings. There were accusations of favouritism and sell-out to the establishment. In retrospect, the detractors seem to have been politically motivated rather than taking into account the best interests of the Stock Exchange and the City itself. Reform was long overdue. The Stock Exchange had been operating in an environment of protected isolation which was now hopelessly outdated. It only remained to see whether it could now remain competitive in a deregulated world, or whether the private fears of the Bank of England would be realised.

Whether all the Stock Exchange member firms were ready for

In September 1983 Sir Nicholas Goodison, Chairman of the London Stock Exchange, suggested the gradual elimination of fixed rate commissions change is open to doubt. The time-table leading towards deregulation was little more than three years. Sir Nicholas Goodison had privately requested a five-year time-table. The smaller broking firms felt that they had been abandoned by the Stock Exchange Council, especially when they saw the more powerful houses being quickly absorbed by major financial institutions at prices which would have been unthinkable a few years before. In September 1983 Sir Nicholas Goodison presented member firms with some suggestions for the gradual elimination of fixed-rate commissions. These were:

- to introduce negotiated rates by class of security (e.g., overseas securities, traded options, gilts, etc.);
- to follow the US example, where commissions on larger bargains were reduced in stages, phased by the size of the transaction, until all commissions were at discretion;
- to introduce upper and lower bands of negotiated commission, leaving a central band of fixed rates, and to increase upper and lower band widths in stages until they met;
- to provide a gradual annual reduction of scales throughout the period to December 1986;
- to give permission to offer discounts against the present rate, the percentage discounts increasing in three stages to full discretion; and
- to provide a specific date on which negotiated rates of commission across the board would apply.

After considerable debate, it was finally agreed that commission rates on overseas securities could be negotiated from April 1984 and that at the same time commission scales on certain gilt-edged bargains would be reduced. While the July 1983 agreement between Cecil Parkinson and the Chairman of the Stock Exchange had assumed the continuation of separate roles between jobbers and brokers, it was quickly realised that this was totally impractical in a practitioner's market. If institutional buyers found that they could deal in gilt-edged securities on a negotiated basis, but had to pay fixed-rate commissions in the equity sector, they would demand zero rates in gilts to compensate for the fixed cost of dealing in equities. At this point it was clearly seen that single capacity was doomed. The only alternative was to remove minimum commissions on all securities on a specified date. The Conservative government, the Bank of England and the Stock Exchange concurred, and a date was agreed. This was later to be confirmed as Monday, 27 October 1986.

After Big Bang

The three-and-a-quarter-year period from the agreement between Cecil Parkinson and the Chairman of the London Stock Exchange will rank as one of the most eventful in the Exchange's long history. After 27 October 1986 the former Stock Exchange structure will be unrecognisable. Many famous names will have been absorbed into anonymous new financial giants. Trading will have moved substantially away from the Stock Exchange floor to a 'Euro' telephone dealing system. Will the Stock Exchange have benefited from all the changes? The answer is unquestionably yes, but some commentators have argued convincingly that the rescue came only just in time. The hundreds of millions of pounds which the new financial institutions have poured into communications systems have made London into one of the most modern financial centres in the world. The series of mergers has resulted in a streamlined structure which replaces a fragmented and often inefficient entity. Trading volume will probably treble within two years. In reasonable market conditions, British securities should find a much wider investor audience. Most London Stock Exchange members will probably miss the old environment, but it had long outlived its usefulness. They may find some consolation in the fact that, following the Parkinson–Goodison agreement, around 750 of the members became millionaires. A statue of Cecil Parkinson on the now much less busy floor of the Stock Exchange would seem the least that they could provide.

CHAPTER TWO

The London equity and gilt-edged markets

The origins of the London Stock Exchange have been described briefly in Chapter 1. Its progression towards becoming a major capital market was assisted by the strength of sterling and the acceptance of sterling as the principal currency for financing world trade. Until the First World War London was the pre-emminent financial centre in the world. Today it ranks as the third capital market behind New York and Tokyo. This performance is still extremely creditable, taking into account two World Wars and the persistent decline in the value of sterling. The rationale for non-resident institutions to retain sterling-denominated assets has not been very strong. There are many bankers today who can remember when there were 25 Swiss francs to the pound. The current rate is around 2.5 Swiss francs, or 10 per cent of the level in the 1930s.

London's success was based on the fact that it recognised the demise of sterling and moved quickly to establish specialist finance activities which would maintain the City's international reputation. Insurance and fund management operations thrived. In both sectors, London could still claim to be the world leader, particularly if the description 'fund management' is extended to 'international fund management' and includes the assets managed by Scottish institutions. Long recognised for their financial skills, the Scottish managers were among the first to identify the investment opportunities abroad. They bought US and Canadian shares at the turn of the century, and were later among the first to move into the rewarding Far East market. They have excelled in the field of venture capital, which essentially forms the basis of the Stock Exchange as a capital raising mechanism.

It is usually assumed that the first securities offered to the public in the United Kingdom constituted government bonds. The first Exchequer stock was sold in 1672. However, almost 100 years earlier the Muscovy Company had sponsored Sebastian Cabot's failed attempt to find the North West Passage. A sum of £6,000 had been raised to finance the three-ship expedition. The East India Company also preceded the sale of the first gilt-edged securities by more than 100 years. The official formation of a company was, in reality, only a successor to the previous system, whereby wealthy individuals or groups provided the financial backing for the earliest exploration voyages. This was venture capital in its truest sense. Returns of 1,000 per cent had been achieved, but the more frequent outcome was a return of zero on a lost ship. The oldest company listed on the Stock Exchange is the New River Company, which was the original 'The Governor and Company of the New River brought from Chadwell and Amwell to London'. This was founded in 1609 by the Earl of Bedford in an ambitious plan to drain the Fens. In 1697 an Act of Parliament was necessary to regulate the activity of brokers and jobbers. By 1845, there were 12 regional Stock Exchanges.

Although London is a distant third behind New York and Tokyo, in terms of the market capitalisation of all securities listed on the Stock Exchange, the statistics are still impressive. According to the London Stock Exchange, the total market capitalisation of all UK equities had reached £292.8 billion as of March 1986, a rise of 19.6 per cent over the December 1985 figure and a gain of 35.4 per cent over the level of a year earlier. Turnover, a vital consideration in the deregulated market, was also higher, rising to £43.3 billion in March 1986 versus £32 billion in December 1985 and £26.6 billion a year before. Gilt-edged turnover, a key consideration for the new primary dealers, showed a rise to £98.1 billion, an improvement of 67 per cent over the previous quarter. Smaller companies coming to the Unlisted Securities Market had a total market capitalisation of £3.89 billion and numbered almost 350.

The equity markets

In the official guide to the City of London the essential function of the Stock Exchange is seen as 'to provide that market-place where buyers and sellers are brought together in a highly competitive atmosphere, in which the day to day price of securities is determined by the simple workings of supply and demand'. This may be true, but it is also an over-simplification of a much more sophisticated mechanism. In fact, the London Stock Exchange has two main functions:

1 as a primary market raising money for governments, government agencies and corporations, both domestic and foreign; and
2 as a secondary market where securities can be bought or sold by domestic and international financial institutions and private clients.

Although London's importance as a financial centre has risen beyond all reasonable expectations, this has been principally due to the emergence of the Euromarkets, to the growth of the foreign exchange market where turnover was recently estimated at £90 billion daily, and to the London International Financial Futures

Nominal and market value of all listed securities at 30 June 1986

	No of Securities	Valuation Nominal £m	Market £m
Public sector: UK & Ireland			
Short	35	39,107.1	39,888.3
Medium	45	53,512.5	60,610.0
Others	42	35,117.0	35,970.3
Subtotal British Funds etc.	122	127,736.6	136,468.6
Short	33	4,005.2	4,035.0
Medium	24	3,481.0	3,579.7
Others	33	1,285.3	1,350.8
Subtotal Irish government	90	8,771.5	8,965.5
Corporation and county stocks – Great Britain & Northern Ireland	157	978.6	982.1
Public boards etc. – Great Britain & Northern Ireland	79	181.3	142.6
Public sector: overseas			
Commonwealth & provincial securities	13	23.0	20.4
Commonwealth corp stocks	2	2.3	1.5
Foreign stocks bonds etc.	118	2,642.5	2,702.6
Corporation stocks: foreign	24	30.8	10.9
Subtotal public sector	**605**	**140,366.6**	**149,294.2**
Eurobonds			
UK companies	107	5,612.8	4,765.3
Overseas companies	1,120	74,615.4	78,081.0
Irish companies	7	405.7	417.5
Subtotal Eurobonds	**1,234**	**80,633.9**	**83,263.8**
Company securities			
Banks & discount houses	109	9,640.3	63,852.6
Breweries & distillers	199	2,224.5	18,822.2
Building societies	16	320.0	321.4
Commercial, industrial	2,727	43,787.1	638,620.6
Financial trusts, land	463	1,748.1	32,050.0
Insurance	59	1,450.4	46,150.4
Investment & unit trusts	496	3,369.6	24,252.1
Mines	143	1,181.4	20,312.1
Oil	66	5,505.8	111,579.5
Plantations	49	381.3	1,264.8
Property	281	2,449.9	11,806.6
Railways	20	171.7	6,012.2
Shipping	28	512.7	3,360.1
Utilities	44	3,463.9	90,665.0
Waterworks	318	415.5	367.9
Subtotal company securities	**5,018**	**76,622.2**	**1,060,437.5**
Overall total	**6,857**	**297,622.7**	**1,301,995.5**

The London Stock Exchange finds itself, perhaps by default, as the uncrowned king of Europe's new securities markets

Exchange (LIFFE). By comparison, London's role as an equity market, although still the largest in Europe, has been steadily declining.

In the Report of the Committee on Invisible Exports, chaired by Sir Thomas Bland and published in 1967, the authors could claim that the London Stock Exchange 'is believed to be the largest exchange, second only to New York'. In that year, turnover on the London Stock Exchange totalled £38 billion. Of this amount some £32 billion was accounted for by trading in gilt-edged stocks and £6 billion in equities. New issue volume, including both fixed interest and equity financing, amounted to no more than £600–£700 million annually. In 1967, the London Stock Exchange had 3,300 members. Even though markets were beginning to recover strongly after the 1973 Yom Kippur war and the subsequent spiral in inflation, there were still only 3,979 members in 1976, and 5,009 members 10 years later.

In 1986 there were 207 stockbroking firms and 18 jobbing houses. Although the London Stock Exchange has a higher membership than, say, New York, its turnover and the combined capitalisation of the listed securities lags well behind not only New York, but also Tokyo, and London's once dominant role in Europe is now being challenged by West Germany.

The need for a major restructuring of the London Stock Exchange is now recognised as being long overdue. But it was only a few years ago that the cocooned European exchanges were seen as indestructible. In Paris stockbrokers were not even allowed to display the name of their firm on the outside of a building. In his book *Men and Money* Paul Ferris wrote 'There is no such thing as a European capital market, which is why so much has been written about it'. Admittedly, those comments were made before 1970. In retrospect, however, they would still have been accurate in 1980. The European stock exchanges had a particularly poor reputation for professionalism and ability to deal in sizeable amounts. Investment analysis was almost non-existent. Insider trading was rampant. By these standards, London was an example of model efficiency, and had it not been for the weakness of sterling and the policies of successive Labour governments, London might have followed the deregulatory path set by New York in 1975. Now the gap between London and Tokyo is far too large to close, and as London's domestic equity markets are simply too small to provide adequate profitability for all the new participating financial institutions, the Stock Exchange has no option but to promote its international role.

Now that markets everywhere are starting to dismantle financial barriers, and the London Stock Exchange finds itself, perhaps by default, as the uncrowned king of Europe's new securities markets, the visions of some far-sighted banking specialists are being realised. As long ago as 1958 the Treaty of Rome, which created the Common Market, saw the need to remove barriers which restricted the flows of international capital between countries. Article 67 of the Treaty of Rome states:

> Member states shall, in the course of the transitional period and to the extent necessary for the proper functioning of the Common Market, progressively abolish as between themselves restriction on the movement of capital belonging to persons resident in member states, also any discriminatory treatment based on the nationality or place of residence of the parties or on the place in which such capital is invested.

The advice of bureaucrats and their banking advisers was not always heeded. London, in particular, was one of the worst offenders. Exchange controls between the members of the original Common Market were dismantled or eased. Only the British held out until 1979, although the premium which they were required to pay in order to purchase overseas securities had been declining since sterling survived the 1976 crisis and moved steadily higher as

Year	Number of Stock Exchange members
1967	3,300
1976	3,979
1977	3,907
1978	3,948
1979	4,028
1980	4,067
1981	4,099
1982	4,116
1983	4,162
1984	4,269
1985	4,495
1986	5,009

In the field of Euro-equities, London has a perfect opportunity to establish itself as the market leader

Nominal and market valuations - British funds

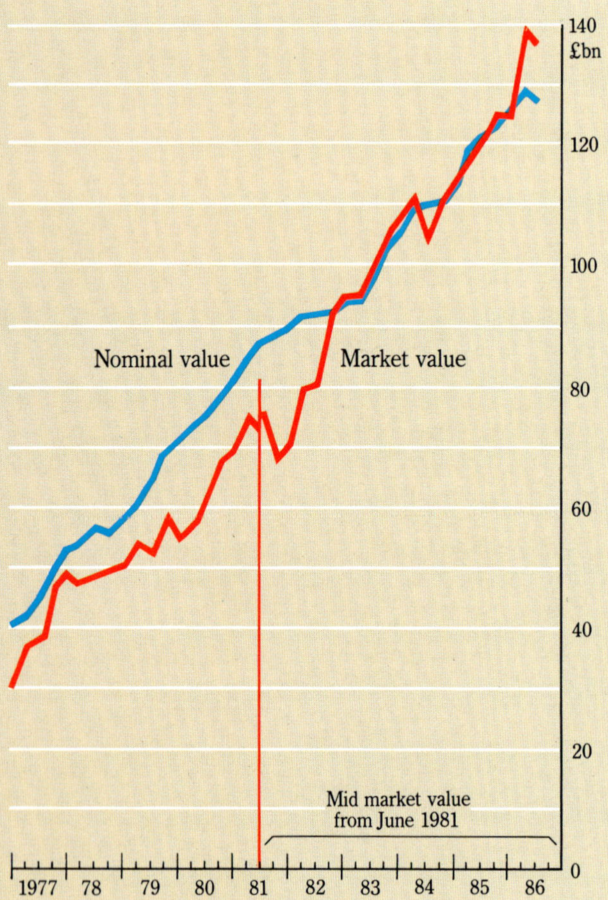

Source: *The Stock Exchange Quarterly,* Summer 1986.

Financial Times ordinary share index

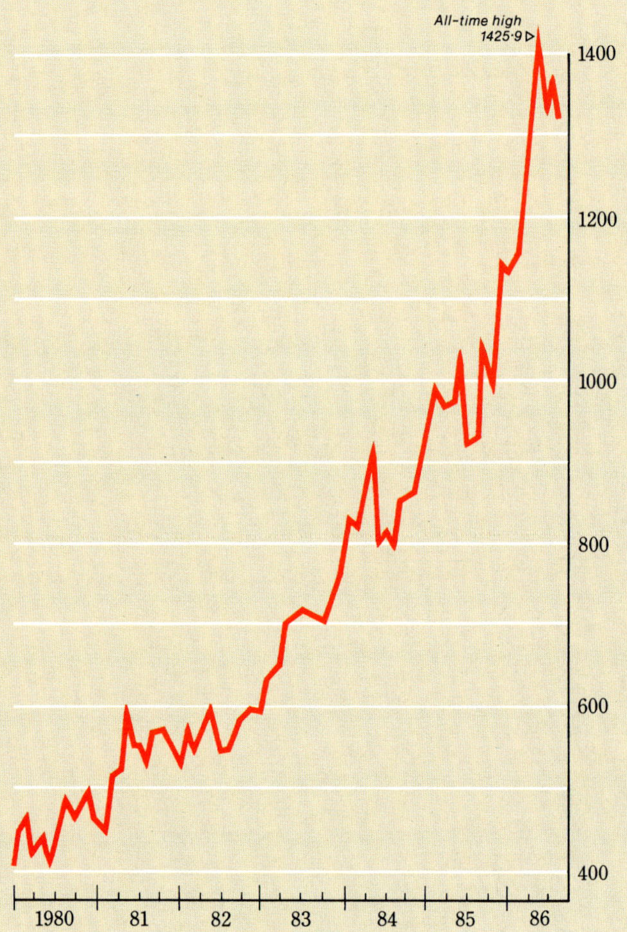

Financial Times Government securities index

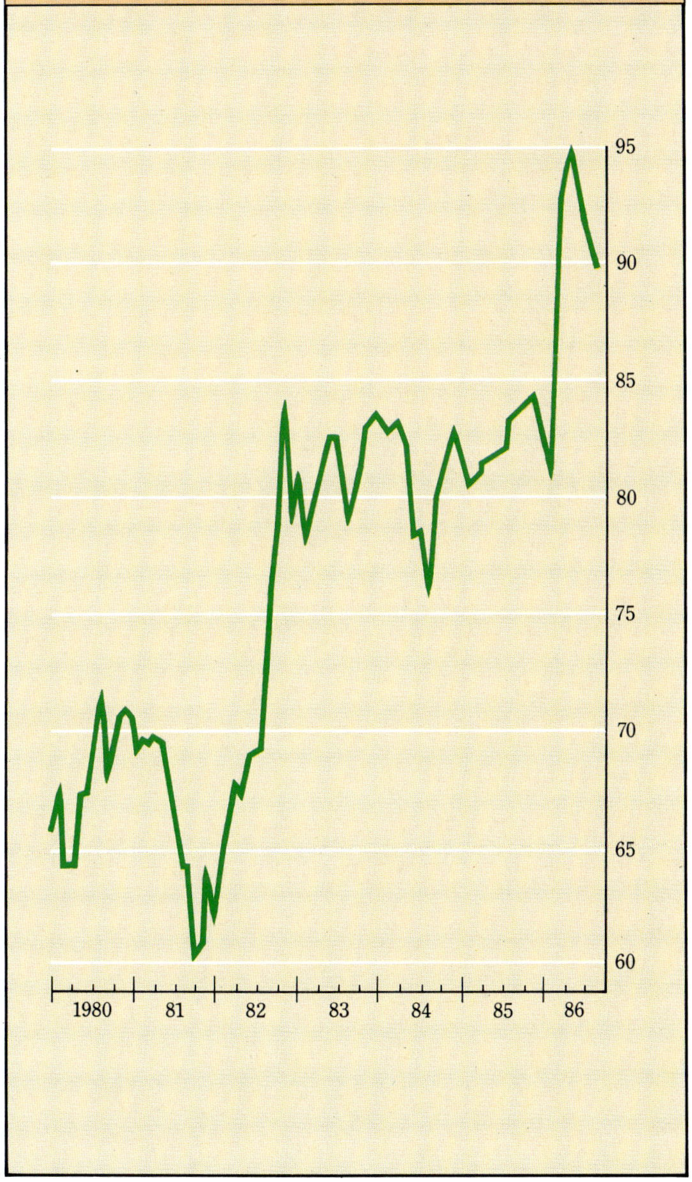

rising North Sea oil production gave sterling the reputation of being a petro-currency.

Big Bang will provide the equity markets in London with a totally new impetus. The larger UK alpha companies will be able to raise huge sums of fresh capital with greater ease and efficiency. The beta companies will now have the opportunity, through the distribution net-works of the foreign institutions, to expand their investor audience beyond the United Kingdom. The gamma stocks will provide niche opportunities for specialist firms. The success of the Unlisted Securities Market (USM) augurs well for smaller companies wishing to go public. The costs of issuing are low and only 25 per cent of the shares have to be offered on the open market. Since 1980 almost 350 companies have listed their shares on the USM.

Euro-equities

In the field of Euro-equities, London has a perfect opportunity to establish itself as the market leader, even accepting the fact that the price may be determined by the stock exchange of the company's country of origin. In European equities, for instance, London has the advantage of longer opening hours and almost permanent access to the largest investing institutions in the world. The London houses must be able to offer Euro-equities in blocks of sufficient size to appeal to the major institutions. This situation would not be dissimilar to the success of American Depository Receipts (ADRs) in New York, where US institutions found that they not only had a time convenience, but also that they could sometimes deal more cheaply and in larger size than by placing orders with London stockbrokers.

The future growth of the Euro-equity market, albeit doubted by some, is in fact one of the Stock Exchange's greatest opportunities in years. London's investment analysts are superior to their Continental European counterparts. The perennial fear that superior analytical capabilities are still inferior to good inside information has lessened as more foreign firms become members of overseas exchanges and establish their own communications systems. Because London will be one of the three recognised financial centres of the world, major companies seeking overseas listings will have little option but to include all three centres. The challenge for London will then be to make an active market in the

Classification of listed securities of companies at 30 June 1986

United Kingdom registered

Section	No. of companies	No. of securities	Loan capital* Nominal amount £m	Market valuation £m
Banks & Discount Houses	33	19	701.7	826.8
Breweries & Distilleries	54	122	731.1	742.3
Building Societies	1	16	320.0	321.4
Commercial & Industrial	1,352	638	4,830.1	5,481.3
Financial Trusts, Land	165	35	198.2	228.1
Insurance	34	7	45.8	42.7
Investment & Unit Trusts	218	152	553.1	610.8
Mines	17	5	188.1	243.7
Oil	31	8	164.1	148.4
Plantations	27	4	8.7	8.2
Property	132	154	1,095.5	1,192.1
Railways	14	11	18.0	4.3
Shipping	16	2	1.4	0.4
Utilities	8	12	13.6	8.2
Waterworks	20	148	176.5	169.1
Total	**2,122**	**1,333**	**9,045.9**	**10,027.8**

Overseas registered

Section	No. of companies	No. of securities	Loan capital* Nominal amount £m	Market valuation £m
Banks & Discount Houses	33	—	—	—
Breweries & Distilleries	2	—	—	—
Commercial & Industrial	207	6	48.0	101.0
Financial Trusts, Land	58	3	14.5	13.8
Insurance	13	1	48.5	50.6
Investment & Unit Trusts	18	—	—	—
Mines	106	6	160.1	194.7
Oil	25	1	12.5	7.6
Plantations	7	1	0.7	0.5
Property	5	—	—	—
Railways	5	2	47.7	98.7
Shipping	1	—	—	—
Utilities	25	—	—	—
Total	**505**	**20**	**332.0**	**466.9**

Republic of Ireland registered

Section	No. of companies	No. of securities	Loan capital* Nominal amount £m	Market valuation £m
Banks	5	2	19.1	20.5
Commercial	67	11	14.2	12.5
Total	**72**	**13**	**33.3**	**33.0**

**Including convertibles*

	Preference capital*			Ordinary & deferred	
No. of securities	Nominal amount £m	Market valuation £m	No. of securities	Nominal amount £m	Market valuation £m
23	89.9	102.1	29	2,800.4	16,220.8
41	447.3	454.2	32	1,020.6	11,799.3
—	—	—	—	—	—
595	1,436.0	1,897.9	1,180	19,046.9	192,577.0
†19	71.2	83.9	77	648.4	5,195.2
9	9.6	8.1	25	657.5	17,079.1
114	126.1	79.9	215	2,099.1	14,154.0
3	11.4	5.5	18	144.9	2,839.8
11	64.0	52.5	26	1,409.3	23,007.1
10	12.1	16.8	27	41.7	291.8
30	29.2	50.1	92	1,292.8	8,884.5
1	1.2	0.3	—	—	—
7	60.4	82.2	18	441.6	2,998.2
2	4.1	11.9	6	33.4	44.6
114	199.7	179.5	56	39.3	19.3
979	**2,562.2**	**3,024.9**	**1,801**	**29,675.9**	**295,110.7**
—	—	—	32	5,917.0	45,913.8
2	11.7	4.4	2	13.8	5,822.0
20	98.6	155.2	198	18,066.3	436,725.4
75	8.5	2,853.6	21	788.2	20,969.5
2	0.1	3.7	15	688.9	28,966.2
—	—	—	15	591.3	9,407.4
12	7.7	37.1	99	669.2	16,991.3
—	—	—	20	3,855.9	88,363.9
—	—	—	7	318.1	947.5
—	—	—	5	32.4	1,679.9
3	0.0	7.4	3	104.8	5,901.5
—	—	—	1	9.3	279.3
1	1.7	0.8	23	3,411.1	90,599.5
115	**128.3**	**3,062.2**	**441**	**34,466.3**	**752,567.2**
—	—	—	4	112.2	768.6
35	10.6	6.1	44	236.4	1,664.2
35	**10.6**	**6.1**	**48**	**348.6**	**2,432.8**

**Including convertibles*

†Excluding 233 participating redeemable preference shares with a nominal value of £19.1 million and a market value of £2,705.9 million.

Nominal and market values of equities by country of origin at 30 June 1986

	No. of companies	Nominal value £m	% Alteration on previous quarter	Market value £m	% Alteration on previous quarter
Australia	12	1,499.0	−3.4	9,383.1	+1.1
Bahamas	1	15.3	−1.3	97.8	—
Belgium	2	0.0	—	2,150.0	−1.7
Bermuda	17	244.7	−3.5	3,112.8	+1.6
Brazil	2	0.0	—	216.0	+151.7
Canada	28	65.6	−2.8	28,317.7	+1.6
Cayman Islands	1	1.3	—	416.0	−9.6
Denmark	4	380.6	+124.3	1,812.4	+79.2
Finland	3	25.7	+9.4	108.3	+72.5
France	4	350.9	+12.1	4,343.5	+9.0
Germany	8	3,538.7	+3.4	30,571.0	−3.7
Hong Kong	2	817.5	+20.6	2,123.3	+3.3
India	2	7.6	−3.8	8.8	+2.3
Israel	3	1.3	—	170.0	+23.0
Italy	1	22.4	—	97.2	−21.6
Japan	8	1,504.7	+5.7	29,479.4	+11.4
Kenya	1	0.1	—	0.4	—
Liberia	1	9.3	+27.4	279.3	+117.0
Luxembourg	8	95.7	−3.2	1,103.2	+9.3
Malaysia	16	457.5	−1.7	1,530.7	+20.0
Malta	1	1.0	—	—	—
Netherlands	15	2,489.4	+1.3	33,966.4	+4.4
Netherlands Antilles	2	13.8	−3.5	6,755.5	+7.1
New Zealand	2	151.6	+6.5	890.1	+19.6
Norway	3	263.1	−6.9	1,645.1	−10.4
Panama	6	21.4	−3.2	483.7	+20.4
Singapore	1	76.1	−2.1	50.4	—
South Africa	94	362.2	−19.6	17,080.2	−27.2
Spain	4	2,359.4	+16.3	6,753.3	+48.8
Sweden	16	849.2	+2.4	8,275.1	+7.2
Switzerland	1	3.7	+5.7	—	—
Taiwan	3	0.0	—	81.2	−1.6
Turkey	1	5.0	—	55.0	+4.8
USA	195	18,778.6	+0.1	561,175.6	+9.0
West Indies	1	0.8	—	—	—
Zambia	2	32.5	−5.0	24.4	−1.2
Zimbabwe	5	20.6	−8.0	10.3	+5.1
Totals	**476**	**34,466.3**	**+2.4**	**752,567.2**	**+7.0**

The new gilt-edged market beyond 1986 is, probably, the most sensitive of all, both in terms of pride and profit expectation

newly listed shares and, with 60 or more market makers, everyone will be trying to identify additional profit opportunities. Previously a London listing was largely an academic exercise, and this must be blamed, at least to some extent, on the inadequacies of the former jobbing system. Overseas listing in 1986 and beyond must mean much more than the ability of a visiting executive to see the company's shares quoted in the country's principal financial newspapers. If there is no accompanying trading in the shares, the listing exercise becomes a waste of time and money. London's new financial markets have a unique opportunity to ensure that they secure a share of this business.

The gilt-edged market

The London gilt-edged market may not be as large as its government securities counterparts in Tokyo and New York, but it has been a model for everyone else for its expertise and ability to raise huge sums of money to fund the public sector borrowing requirement with apparent ease, even in the most difficult times.

Prior to recognition of equity values, which did not begin in earnest until the late 1950s, gilts were the standard barometer of the health of the market. Their importance has, perhaps, been forgotten by subsequent generations whose fathers and grandfathers used to gauge the prevailing sentiment of the market by watching the price of Consols. In the book *The Post War History of the Stock Market* the gilt-edged Consols are given a whole chapter, and it is made quite clear how important they were considered to be as an indicator. The following excerpt describes the influence of Consols between 1956 and 1957.

> The decline of Consols was not interrupted until the end of 1956 when a substantial rally, which consisted of a single large tertiary movement, lifted Consols from 50 to 56. This rally accompanied a strong upward movement of the index from the oversold position that had developed after the Suez crisis. The rise of Consols lent considerable respectability to the early phase of this rise, but by February Consols had flagged and by March the trend was again downward. The message that should have been found by the bulls of shares, if they had studied the curve of Consols, was: 'This is a calf market; do not trust it too far.'
>
> The bear market of 1955–58 did not, like that of 1951–52, contain a single crisis but rather, like the bear market of 1947–49, a pair of twins. Consols were warning in the first half of 1957, as in the second half of 1947, that there was more trouble to come; and the warning was, if we may judge by the curves of activity, much less heeded.

Today the language, admittedly, sounds ponderous and it is necessary to read the whole of the book to understand fully the meanings of 'activity curves' and 'calf' markets. However, while the efficiency of the gilt-edged market was never in doubt, there was already a growing recognition that there was a potential fixed-income market beyond gilts. The Report of the Committee on Invisible Exports, chaired by Sir Thomas Bland, saw not only a wider capital market, but was also mildly critical of the Stock Exchange itself.

> While it is clear that any hopes of increasing sterling issues for foreign borrowers in London must await a significant improvement in the United Kingdom's balance of payments, the expansion of international issues, denominated in foreign currencies and subscribed to by non-residents, needs further encouragement. The committee believes that every help should be given to enable London to maintain its position as the main entrepot centre for international operations of this kind. Although the market continues to grow rapidly, competition is increasing and London appears to be hampered by a number of what appear to be unnecessary restrictions.
>
> It was represented to the Committee that the Stock Exchange's Certificate of Exemption from the requirements of the Companies Act in practice limits the exemption only to those sections of the Act which are unduly burdensome for the borrower. Sections which are virtually meaningless in the context of a foreign loan, it was complained, are nevertheless enforced. The borrower is compelled to go through unnecessary and irritating technicalities. The Stock Exchange Council by a simple directive, it was suggested, could assist invisible earnings.

The Bland report is now almost 20 years old, but its board members and support staff should be given credit for having the vision to see the changes which lay ahead. The new gilt-edged market beyond 1986 is probably the most sensitive of all, both in terms of pride and profit expectation. Since the Bank of England invited applications for primary market gilt dealerships, there have been four drop-outs from the original accepted list of 31 names. The earliest were J. Henry Schroder Wagg and Drexel Burnham Lambert. Since that time Bank of America has withdrawn, and most recently, and perhaps surprisingly, Union Discount Co.

Initially there will be an excess of bravado, an urgency to buy business at almost any price and probably a virtual absence of profits

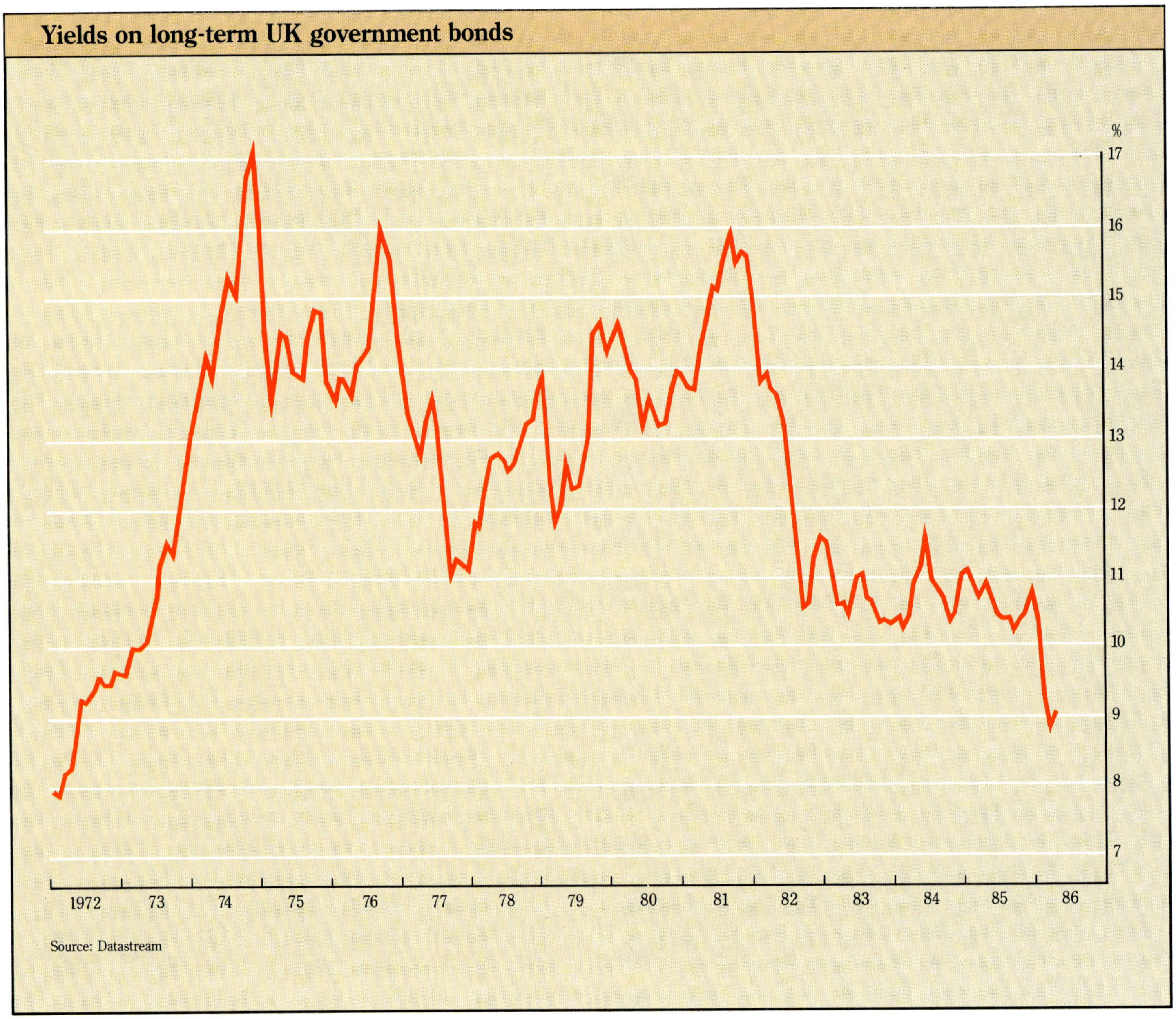

Stock Exchange turnover of UK government securities

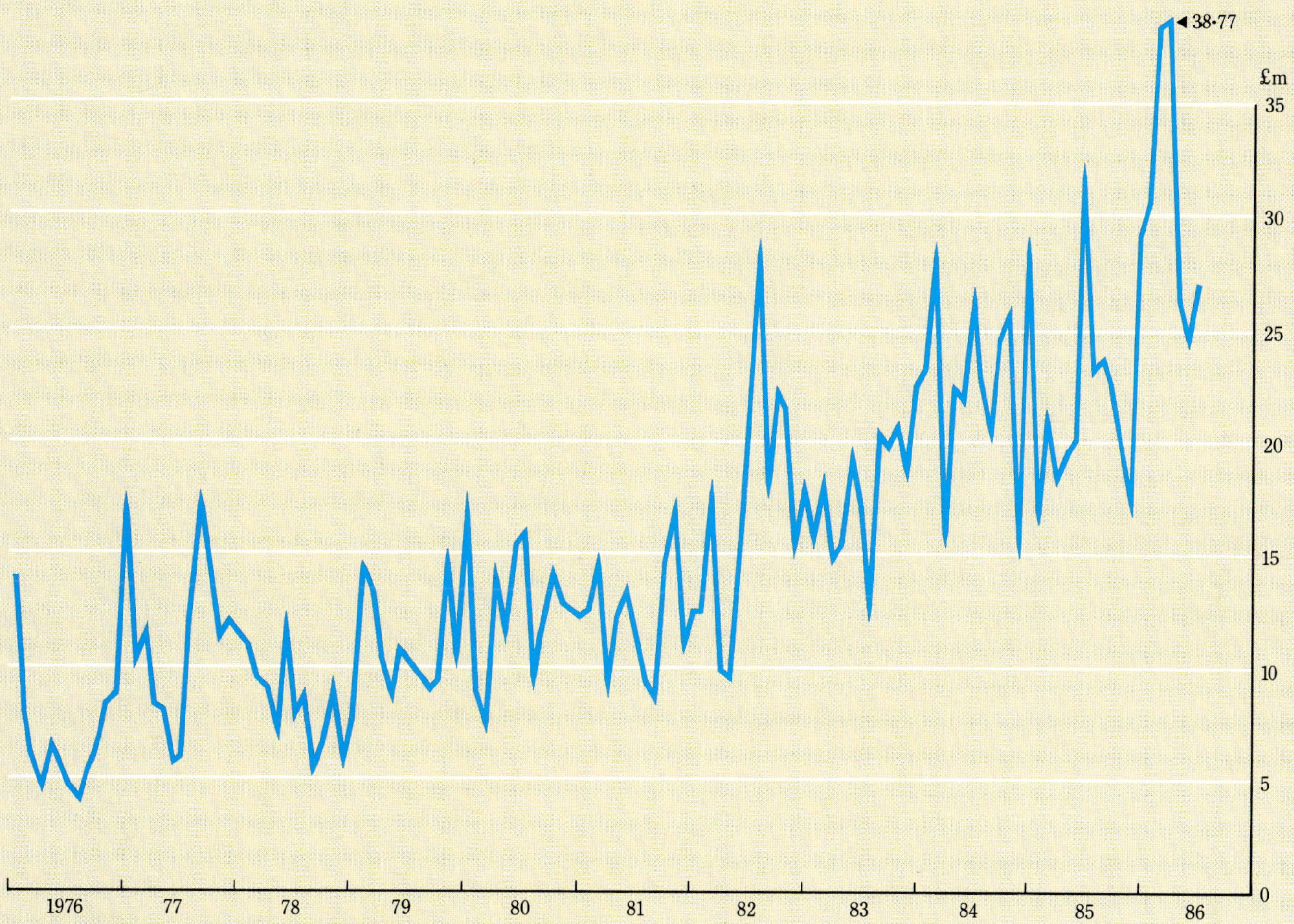

Source: Datastream

Initially there will be an excess of bravado, an urgency to buy business at almost any price and probably a virtual absence of profit. Whereas the intended participants were only too eager to discuss their anticipated market share in late 1985 and early 1986, the mood is now much more reserved and clearly some former predictions are now being seen as hopelessly optimistic. The suggestion that there will be a substantial number of further withdrawals from the gilt market is discussed in later chapters and is not accepted. For most houses the die is cast. The numbers may contract through mergers, but there will also be additions. Profitability is not an immediate necessity for a group of houses which comprise some of the most powerful financial institutions in the world. In most cases gilts are not even a separate component, but part of a global fixed-income trading group. Gilt losses, if there are losses, will be offset by gains in futures, foreign currency bonds and other instruments. A year ago, when UK gilt specialists were saying that there was only sufficient business to support around a dozen dealers, they may not have taken into account that gilts were only a part of a much grander and more ambitious long-term strategy.

Many houses have, however, incurred substantial start-up costs, which will make profits hard to earn and break-even a more reasonable expectation for some. Dealing costs will be drastically reduced. In a recent book *How the New City Works*, Colin Chapman comments 'It is also feared that the costs of dealing are going to rise, not least because of the salaries being paid and the carving up of the market among a greater number than under the old system'. This would appear to be a contradiction: greater competition has rarely resulted in higher costs unless the system is highly inefficient. The salaries being paid are only the market price for the job.

In essence, there have been two major gilt-jobbers, Akroyd and Smithers and Wedd Durlacher. When these two firms were absorbed by Mercury International and Barclays Merchant Bank, there was a rule for the remainder. The salaries being paid will have to be earned and, if there is a short-fall, there will be a number of redundancies. Perhaps more UK firms should have followed the US example and trained their own gilt dealing staff according to the American attitude that a good trader can make markets in anything, whether it be US Treasuries, Euros or gilts.

The new gilt market will be highly competitive, risky and a natural breeding ground for a stream of new financial innovations. It will be much less gentlemanly as the market moves away from the traditional trading floor to a fast-moving telephone system. The transition towards Euro-style trading will be increased by the arrival of the interdealer brokers who, as in the case of the Eurobond market, can only deal between primary market makers, and who may trade up to 200 times their capital each day. The brokers expect to handle a substantial amount of the daily volume in gilts, and estimates of around 50 per cent may not be totally unrealistic if the US houses capture as high a market share as expected. In the United States the brokers are an integral part of the US Treasury market.

If the 27 primary market gilt dealers are all beginning to trade in an environment of a bear market in both gilts and sterling, the signs are ominous. This may not be the case. In that case the overseas buyers of gilts would be the first to withdraw, which would leave the British houses, with their long established UK institutional connections, at an advantage. However, houses such as Merrill Lynch have already proved that client relationships are not sacrosanct, and the US houses, with their formidable innovative skills developed in the Euromarkets, will be introducing new products, which will radically alter the traditional image of the gilt-edged sector. In a recent article prepared in a special *Euromoney* supplement, the commentary on the new gilts market was entitled 'Goodbye Cosy World'. How correct this is, and one wonders whether the old Consols-watchers would ever have appreciated that their principal market barometer would, one day, be just another characterless security being passed around the world in a 24-hour global trading system.

The London Stock Exchange: 6 Capel Court entrance

CHAPTER THREE

London and the Euromarkets

In many respects London is the Euromarket. Perhaps this would be the ultimate tribute to the late Sir Siegmund Warburg who developed the market here but in the initial stages saw his efforts soundly rejected by a short-sighted City community. London, as a financial centre, has much to commend it, as Wood Gundy's Ian Steers is quoted in the Introduction. However, there was no special reason why London should have developed the Euromarkets into a near monopoly. Paris was considered to be one of the alternative forerunners, and in the 1960s some of the most important primary market houses, such as Morgan & Cie (now Morgan Stanley International) and Lehman Brothers International (now Shearson Lehman Brothers International) had their headquarters based in Paris. Amsterdam was another contender. Luxembourg would have dearly liked to have housed the Eurobond market and made its own invaluable contribution by facilitating the listing of Eurosecurities both efficiently and cheaply. The Swiss cities of Geneva and Zurich were logical centres as, until very recently, Swiss banks were by far the largest buyers of Eurobonds. The drawbacks seem to have been the presence of turnover tax, which would have eroded dealers' profitability, and the reluctance of the Swiss authorities to permit entry of the large number of foreign personnel required to operate the market efficiently. The two main clearing houses for international securities, Euroclear SA and Cedel SA, are both based on the Continent rather than in London.

The gravitation towards London was initially a slow process. The Japanese securities houses came to London with the strict objective of selling equities. That too was the intention of the major US houses. Salomon Brothers opened a London office in January 1971. However, 'by 1973 we had realised that the Eurodollar business was the one to be in', said Charles McVeigh, Salomon's Senior Managing Director in London. The Japanese, European and US banks, which mushroomed in the City, existed for traditional commercial banking business. No doubt they could see the burgeoning Euromarkets securities business, but it took a long time for them to decide to become participants. Deutsche Bank, always one of the most powerful forces in the Euromarkets, only moved part of its capital markets and trading group to London in 1984. UBS (Securities) Limited, although a subsidiary of Union Bank of Switzerland, probably the most consistent of all Eurobond buyers, had a London staff of only 25 until 1984. Many European banks still retain their capital market and trading groups in their country of origin.

A missed opportunity

Even if domestic pride sometimes inhibited the growth of the overseas institutional offices in London, it was the British themselves who again failed to seize the opportunity. Admittedly, the UK merchant banks increased their commitment to the market by expanding secondary market trading capabilities and adopting a more aggressive attitude towards winning primary market mandates. However, the record shows that only S. G. Warburg &

Salomon Brothers' Charles McVeigh: 'By 1973 we had realised that the Eurodollar business was the one to be in'

Co has consistently maintained a prominent position in the Eurobond league tables, and that in the secondary market UK merchant banks are not seen to be in the premier category. Only in Eurosterling bonds are they considered to be an effective force.

It is difficult to see why the UK merchant banks offered no real opposition to their foreign competitors. Why was Credit Suisse First Boston allowed to dominate the primary markets for so long, and even if one accepts that CSFB is a very special organisation, why were CSFB's main competitors Morgan Stanley International, Salomon Brothers International, UBS (Securities), Swiss Bank Corporation International and Deutsche Bank? As if this was not a sad indictment of the City of London, there is a new presence in the markets, which constitutes perhaps the major threat to the traditional Eurobond leaders, and which will move most UK merchant banks further down the league tables.

The new threat comes from the Japanese banks and securities houses, who make no attempt to disguise their ambition to be at the top of the very visible, if not the most profitable, Eurobond primary market league tables. The Japanese come with the most impressive credentials. They are the largest investors in overseas securities in the world. To date, these capital flows have been directed mainly into bonds, with equity purchases contributing less than 5 per cent. As the largest buyers in the world, now even outranking the Swiss banks, they are a natural target for any borrowers wishing to diversify their traditional sources of funding. Japanese bank and securities houses, therefore, have had no difficulty whatsoever in catapulting themselves up the league tables. There have been the usual accusations of 'buying' issues and suicidal bidding, but the fact remains that they can buy their way into the top of any capital market they wish and have already become a major force in the Eurobond market. If the buying power of the Japanese institutions is not sufficiently awesome in itself, they have other natural advantages. Japanese institutions are not only the largest net foreign buyers in the world, Japanese companies and banks are also among the largest borrowers in the world. This combination becomes an irresistible force when taking into account that Japanese borrowers use, whenever possible, Japanese lead managers. In 1986 Nomura International has been lead manager for more individual issues than Credit Suisse First Boston, although CSFB is still well ahead in the total volume of capital which the issues represent.

The sheer aggression of the US and European houses and the resources of the Japanese financial institutions do not bode well for the UK houses in the Euromarkets. Many have said openly that they have no intention of taking on the cash-rich newcomers in a head-on confrontation. Rupert Hambro, the former Chairman of Hambros Bank plc, was so certain that it would be difficult to compete that he and various family members and colleagues left the bank to form a new organisation which will concentrate on venture capital and specialist strategic holdings. 'We decided that it was time to go back to basics,' said Rupert Hambro. The Hon. David Poole of ANZ Merchant Bank was even more pessimistic: 'UK houses won't stand a chance versus the Salomons of this world. Never take on competition that you can't beat.'

The most likely outcome for London houses in the Euromarkets is that they will leave the foreign competitors with their present near monopoly. In the period following deregulation, the domestic UK merchant banks will be too busy attempting to retain their market share in gilts and London equities to consider seriously taking market share in the Euro sector away from companies such as CSFB, Morgan Guaranty or Goldman Sachs. Indeed, the pressure may be in reverse. Foreign banks have already proved, beyond any doubt, that they can win mandates for UK borrowers from the domestic merchant banks. Credit Suisse First Boston achieved the ultimate distinction when it was selected as book-runner for the US$2.5 billion seven-year floating rate note

The Hon. David Poole of ANZ Merchant Bank: 'Never take on competition you can't beat'

Many UK houses have no intention of taking on the cash-rich newcomers in a head-on confrontation

issue for the United Kingdom in October 1985. This was the largest public Euro offering in the market's history and although S. G. Warburg & Co appeared in the prized top left-hand position, the books were run by CSFB, who were therefore given recognition for the mandate.

The award of the United Kingdom mandate to CSFB was not an isolated example of the inroads being made by foreign banks into what had once been the exclusive territory of the UK merchant banks. In November 1985, BTR plc, one of the most successful UK industrial companies, issued US$150 million and ECU170 million of convertible debentures with Swiss Bank Corporation International as the lead manager. The structure of the loan, which featured low interest coupons and a low conversion premium with a put option at the choice of the bondholder, was a thinly disguised 'rights' issue which by-passed the London Stock Exchange. Lonrho plc and the Hawley Group both issued Euroconvertibles soon afterwards, using Credit Suisse First Boston as the lead manager. If, as suspected, more and more top-rated UK industrial companies by-pass the traditional 'rights' issue mechanism and instead use the Euro alternative, the chances are high that the overseas banks will win a high proportion of this business.

Perhaps the greatest failure of the London Stock Exchange has been not to recognise the winds of change and to jump on the Euro-bandwagon too late. The 1986 register of members of the Association of International Bond Dealers listed 32 UK stock-brokers and jobbers. Membership of the AIBD implies a consistent interest in the Euromarkets. On closer examination of this list, however, one can conclude that there are no more than six who would be considered to have any serious influence in the market, and only one which would rank among the genuine market maker leaders. When it is considered that the market began in London, that the first Eurobonds were traded in London and that the Eurobond market dwarfs the domestic sector, it would seem more than a careless oversight by the London members of the Stock Exchange.

The excuses were that the member firms did not have a client base overseas. Domestic investors were penalised by the existence of UK foreign exchange controls and the necessity to pay the foreign currency premium, of which a part had to be surrendered on resale. The premium was certainly a negative influence, but it should not have been a total disaster, particularly in the case of equity-related Eurobonds. Some of the first Japanese convertible Eurobonds rose in price by almost 1,000 per cent. Capital gains of 200 to 300 per cent were quite common. During this period sterling was also a persistently weak currency. Under the new dual-capacity system, UK resident investors are only dealing in the same way as Strauss Turnbull were transacting business 23 years ago. Stanley Ross, now a Managing Director of Deutsche Bank Capital Markets in London, but a senior dealer for Strauss when the Eurobond sector opened, commented 'we learned to be net traders through our arbitrage activities'. Ross was, nevertheless, disparaging in his views of the London Stock Exchange, which he described as

Ian Steers, Chairman of Wood Gundy: 'Only a major change in government policy would cause the market to move'

Every UK merchant bank and stockbroker is engaged in a frantic scramble to build up trading strengths in Far Eastern securities

'displaying a degree of insularity comparable only to New York in the 1960s and 1970s'.

It remains a mystery, however, why some London Stock Exchange member firms did not emulate Strauss Turnbull's success, particularly when many were perhaps better capitalised and had access to an equally wide range of overseas clients. Many firms provided European equity research at the same time as being a force in the gilt-edged or UK corporate debenture markets. It should have been possible to blend the two. If the actual trading aspect posed psychological problems because of lack of experience, it would have been relatively simple at that time to hire a young dealer from one of the London jobbing firms who would have a natural trading background. Many London Stock Exchange firms are now belatedly looking enviously at the Euromarkets, but they should have acted decisively long ago. When Interest Equalisation Tax was reduced to zero in August 1974 and the market continued to expand in the post-OPEC-driven economic crisis, it should have been evident that its future was assured. Even more recently it was quite possible to gain profitable entry into the Euromarkets through niche participations.

Deutsche Bank Capital Market's Stanley Ross: 'The London Stock Exchange displayed a degree of insularity comparable only to New York in the 1960s and 1970s'

The Euromarkets have grown so quickly with so many new instruments that it is no longer possible for one house to trade them all. If a typical London Stock Exchange firm is in the process of merging with a domestic or foreign banking partner, it may have no direct Euromarkets operations of its own but could conceivably offer areas of research expertise which would complement the bank's existing Euro activities. A classic example of profitable niche participation has been the strategy of Société Générale Strauss Turnbull (SGST) in adapting itself to the changing market environment. Having decided not to compete directly with the giant foreign trading houses in traditional areas such as straight dollar bonds and US convertibles, it chose to become specialists in debt warrants. The timing could not have been better. The debt warrants market mushroomed as lead managers found that they could lower total funding costs and assist the completion of swaps. Conventional debt warrants became detached warrants, 'wedding' warrants, income-producing warrants, and put and call warrants. The bonds into which they were exercisable became 'virgin' bonds. As interest rates tumbled in the period 1985 to mid-1986, the value of some debt warrants increased by 400 to 500 per cent. Although trading any form of warrant, by definition a highly leveraged instrument, involves an above average degree of risk, SGST developed almost a monopoly in debt warrants and at the same time almost certainly made a very substantial profit.

Most of the UK commercial and merchant banks, together with their stockbroking or jobbing subsidiaries, intend to become primary dealers in the gilt-edged market. A logical extension of gilt-edged principal trading is to make prices in Eurosterling bonds. This may prove to be a mistake. To date, the trading of these bonds has been an erratic but generally profitable niche area of the market. However, there is insufficient volume, at least at the present time, to support the presence of more than four or five

PROSPECTUS

U.S. $2,500,000,000

United Kingdom

Floating Rate Notes Due 1992

The issue price of the Notes is 100 per cent. of their principal amount.

Principal of and interest on the Notes will be paid as described under "Description of the Notes—Payments". Payments will be made without deduction for United Kingdom withholding taxes, as described under "Description of the Notes—Taxation".

The Notes will mature on the Interest Payment Date falling in October 1992. The Notes are subject to redemption on and after the Interest Payment Date falling in October 1988, at the option of H.M. Treasury, in whole or in part, at their principal amount. In addition, H.M. Treasury will, at the option of the holder of any Note, redeem, on the Interest Payment Date falling in October 1990, such Note at par.

Interest on the Notes, which will be payable quarterly in arrear in January, April, July and October in each year beginning in January 1986, will (subject to a minimum rate of 5 per cent. per annum) accrue at an annual rate equal to the London interbank bid rate for three month Eurodollar deposits as described under "Description of the Notes—Interest".

Application has been made to the Council of The Stock Exchange of the United Kingdom and the Republic of Ireland ("The Stock Exchange") for the Notes to be admitted to the Official List.

The Notes will initially be represented by a temporary Global Note to be deposited with a common depositary for both CEDEL S.A. and Morgan Guaranty Trust Company of New York, Brussels office, as operator of the Euro-clear System on or about 7th October, 1985. The temporary Global Note will be exchangeable for definitive Notes not earlier than 90 days following completion of the distribution of the Notes, as determined by Credit Suisse First Boston Limited, and upon certification as described under "Description of the Notes—Form and Denomination".

S. G. Warburg & Co. Ltd. **Credit Suisse First Boston Limited**

Amro International Limited **Bank of Tokyo International Limited**
Bankers Trust International Limited **Banque Bruxelles Lambert S.A.** **Banque Nationale de Paris**
Barclays Merchant Bank Limited **Baring Brothers & Co., Limited**
Citicorp Investment Bank Limited **Commerzbank Aktiengesellschaft** **County Bank Limited**
Deutsche Bank Capital Markets Limited **Dresdner Bank Aktiengesellschaft**
Goldman Sachs International Corp. **Hambros Bank Limited** **Hill Samuel & Co. Limited**
Kleinwort, Benson Limited **Lazard Brothers & Co., Limited** **Lloyds Merchant Bank Limited**
Merrill Lynch International & Co. **Samuel Montagu & Co. Limited**
Morgan Grenfell & Co. Limited **Morgan Guaranty Ltd** **Morgan Stanley International**
Nomura International Limited **Orion Royal Bank Limited** **N. M. Rothschild & Sons Limited**
Salomon Brothers International Limited **J. Henry Schroder Wagg & Co. Limited**
Standard Chartered Merchant Bank Limited **Swiss Bank Corporation International Limited**
Union Bank of Switzerland (Securities) Limited

The date of this Prospectus is 23rd September, 1985.

BTR plc

(Incorporated with limited liability in England under the Companies Acts 1862-93)

U.S. $150,000,000
5 per cent. Convertible Subordinated Bonds due 1995

and

ECU 170,000,000
4¼ per cent. Convertible Subordinated Bonds due 1995

both convertible into Ordinary Shares of 25p each of BTR plc

Issue Price for each issue 100 per cent.

The Bonds are convertible into Ordinary shares of 25p each of BTR plc, at 390p per Ordinary Share, subject in each case to adjustment under certain conditions, on or after 1st January, 1986. On 12th November, 1985, the middle market quotation for the Ordinary Shares as derived from The Stock Exchange Daily Official List was 380p.

Application has been made to the Council of The Stock Exchange for the Bonds to be admitted to the Official List.

Temporary Global Bonds, without coupons, will be deposited for the account of the European clearance systems on or about 3rd December, 1985 and will be exchangeable for definitive Bonds of each issue after the expiry of 90 days from the completion of the distribution of the Bonds upon certification of non-United States beneficial ownership.

Swiss Bank Corporation International Limited Hill Samuel & Co. Limited

Credit Suisse First Boston Limited Goldman Sachs International Corp.
Morgan Grenfell & Co. Limited Nomura International Limited
Union Bank of Switzerland (Securities) Limited

Amro International Limited Banque Bruxelles Lambert S.A.
Banque Nationale de Paris Berliner Handels- und Frankfurter Bank
Cazenove & Co. Deutsche Bank Capital Markets Limited
Dresdner Bank Aktiengesellschaft Enskilda Securities – Skandinaviska Enskilda Limited
Generale Bank Kidder, Peabody International Limited
Lazard Frères et Cie Merrill Lynch International & Co.
Morgan Stanley International N.M. Rothschild & Sons Limited
Salomon Brothers International Limited S.G. Warburg & Co. Ltd.

Julius Baer International Limited Banca del Gottardo
Banca della Svizzera Italiana Bank Leu International Ltd
Bank J. Vontobel & Co. AG Compagnie de Banque et d'Investissements, CBI
HandelsBank N.W. (Overseas) Ltd. Lombard Odier International Underwriters S.A.
Pictet International Ltd. Swiss Volksbank

Offering Circular dated 13th November, 1985

serious market makers. If even half of the proposed 27 primary gilt-edged dealers also decide to make a market in Eurosterling bonds there may be no profit for anyone.

The example of Robert Fleming

The secret of future success in the Euromarkets, for those who are not already in the premier league, will be to find areas of the Euromarkets which are either about to expand rapidly or which are underexploited. The combination of new trading skills with existing in-house expertise is an ideal basis. Therefore, when the UK merchant bank Robert Fleming and Co Limited announced that they would begin trading Eurobonds, they had no intention of entering the highly visible but also highly competitive, low margin straight dollar or FRN markets. The first area of concentration was in Japanese convertibles and Japanese equity warrants. Why did the new Robert Fleming Securities Limited choose the Japanese markets? The answer lay in the parent bank's investment management skills and, in particular, Fleming's knowledge of the Far Eastern markets including Japan. In-house research capabilities were therefore already at hand. Excellent trading connections with Tokyo were long established. In 1986, Robert Fleming became one of the first overseas banks to be admitted as a full member of the Tokyo Stock Exchange. Again the market's direction was providential. When Robert Fleming Securities was started in 1983 they may not have foreseen that, in the 1983–86 period, the Tokyo stock markets would be among the world's star performers or that the Japanese yen would move to new highs against the US dollar, which even exceeded those reached in 1978. In the end it is results which count, and the new Robert Fleming securities trading unit was soon making a significant contribution to the profits of the parent bank.

The direction taken by Robert Fleming and its subsequent success is the natural approach for UK houses in the Euromarkets. Fleming simply built on its own in-house strengths. Unfortunately, every UK merchant bank and stockbroker is now engaged in a frantic scramble to build up trading and distribution strengths in Far Eastern securities because this is currently the most fashionable investment area in the world. One must ask why the majority are so late. The London stockbroking house of Vickers da Costa, now part of Citicorp, had been specialists in Japanese equities for more than 20 years. Other London houses, including W. I. Carr, Cazenove and Hoare Govett, have been active in Japanese securities for a similar period. These institutions have established a lead which places them at a huge advantage and which they are unlikely to

CSFB achieved the ultimate distinction of being selected as book-runner for the UK's 1985 US$2.5 billion FRN issue

In 1985 BTR issued convertible debentures with Swiss Bank Corporation International as the lead manager

surrender. The late arrivals can either choose to abandon the chase or adopt an 'open cheque book' policy and hope that the hiring of specialists with proven market expertise will produce a lasting market share.

The Japanese example is classic in that it portrays almost exactly the type of situation which the City will confront in the period following deregulation. There is a huge demand for Japanese equities. There is also a similar level of demand for Japanese convertible bonds and Japanese equity warrants. Will the London houses be able to ensure that they receive a reasonable share of this business?

The omens are less than encouraging. A substantial proportion of the equity will continue to be channelled directly to the Japanese broking houses in return for favoured treatment in new issues and the provision of research information. In addition, the few UK firms which do have full membership of the Tokyo Stock Exchange will find that the competition for Japanese equity business will come not only from Japanese securities houses, but also from other foreign houses, particularly the US investment banks. When Morgan Stanley & Co went public in early 1986, the prospectus clearly stated that part of the proceeds would be used to expand their offices in London and Tokyo. Merrill Lynch & Co have the intention of becoming one of the most significant factors in the distribution of Japanese equities and are currently building up a Tokyo sales, research and trading team which will eventually add several hundred more employees to the existing staff. Salomon Inc, the premier house in US government bond trading, is already one of the six largest dealers in domestic Japanese government bonds.

This is the type of competition which the UK houses will have to face, not only in the City of London, but in every major developed capital market in the world. Their competitors have virtually unlimited capital resources, no respect for existing salary scales or traditional staff loyalty, and are quite prepared to 'buy' short-term market share at the expense of immediate profit. For the London houses now facing that challenge it must seem a daunting proposition.

In the highly volatile but often hugely profitable Japanese convertible and equity warrant Eurobond market, the UK presence, with the exception of houses such as Robert Fleming, is almost non-existent. Many execute orders for clients but the real profits are in trading and positioning. Why should some of the UK houses, with 20 years' experience in the Japanese markets, not have transferred some of that capability to the Eurobond market where Japanese companies are, coincidentally, the largest borrowers when measured by the amount of individual new issues? The transition should have been relatively simple. In the trading of Japanese Euroconvertibles in London, a firm such as Vickers da Costa should have ranked in the same category as Nomura or Daiwa. Even when Robert Fleming began its highly successful Japanese Euro trading operations some years ago, other UK merchant banks were slow to follow, if indeed they followed at all. S. G. Warburg and Co, for instance, would have seemed to be an ideal candidate with their full membership of the Tokyo Stock Exchange and their historical record of lead managing some of the first Japanese convertible Eurobond issues. However, S. G. Warburg & Co has never tried to promote itself as a major secondary market maker in spite of the bank's very substantial participation in almost all sectors of the Japanese market.

Julius Strauss, of Strauss Turnbull: 'The jobbers' loss was our gain'

The late arrivals can either choose to abandon the chase or adopt an open cheque book policy

Is the Japanese Eurobond example another instance of lost opportunity for the City of London? It is fortunate for the City that a high proportion of the trading volume in Japanese convertibles and Japanese equity warrant Eurobonds is still conducted through London. It is unfortunate that only a small part of this lucrative business is being handled by the UK Eurobond houses. The reluctance of the UK Eurobond specialists to participate more actively in this market is difficult to comprehend. 'The Japanese have a natural information advantage over us all, as well as access to cheap yen financing and lower equity commissions', said one UK merchant banker. While these points are entirely valid, they have not deterred companies such as Morgan Stanley International and Merrill Lynch, who not only trade US dollar Japanese Euroconvertibles but also Japanese issues denominated in Swiss francs and Deutschmarks. This should not be seen as direct aggression on the part of the US investment banking houses, but

Charles Villiers, Chief Executive of County Bank: 'US banks are our model'

Michael Hawkes, Chairman of Kleinwort Benson, foresaw the coming of deregulation

The real lesson from the development of a market such as swaps is that imitation can be as successful as innovation

rather they are following the natural progression of Japanese equity-related business. As Japanese convertible and equity-related Eurobonds at least equal the amount of Japanese equity offered directly abroad, it would be inappropriate to specialise in one sector and not the other, particularly if the final securities being offered are essentially the same. It does not necessarily make any difference if the bonds are denominated in Eurodollars, Euro Swiss francs or EuroDeutschmarks. The key determinant is that the bonds represent an entitlement to the underlying Japanese equity.

Morgan Stanley's efforts in building up non-US dollar Japanese Eurobond capability has been impressive. 'We always check against MSI's prices before deciding where to place our Japanese orders and I am talking about issues denominated in Swiss francs', said a prominent portfolio manager based in Zurich. As the entry of the US investment banking houses into the non-US dollar Japanese equity-related bond sector is less than four years old, the UK Eurobond houses should perhaps be discussing how they managed to squander a full 15 years' advantage.

The test ahead

The real test for the UK financial institutions in the period after deregulation could lie in the role which they see themselves playing in the government bond markets of other major developed countries. Their own territory in the UK gilt-edged market has been invaded by foreign competition. However, they in turn can counter by trading, say, German government bonds or Japanese government bonds in London. There is already a very substantial amount of business transacted in these securities through London, and it will be interesting to see if the parent companies of Wedd Durlacher and Akroyd & Smithers, who once dominated the London gilt-edged market, will at some time in the future become a major influence in non-sterling government markets.

As the term 'Euro' becomes increasingly to mean 'international' as markets deregulate, the differentials between government bond markets will lessen in terms of the complexity in executing transactions. To place the sum of, say, US$10 million equivalent in each of the sterling, US dollar, Deutschmark and yen markets previously needed four telephone calls. Today one telephone call to a major international securities house should suffice. If the major international securities house can, in addition provide extra services in the form of research and currency forecasting, it will be at a major advantage over the competition. The much-discussed US objective of 'one-stop shopping' has proved to be attractive only in theory rather than in practice.

To date no single house has been able to provide all the services that a sophisticated professional investor requires. However, the time is rapidly approaching when at least a dozen international securities houses will lay claim to being able to provide worldwide,

CIBC's Jamie Clark has built up enviable niches in the Eurobond market

multinational currency execution capability. A number of the new UK consortia are being deliberately structured on this basis and some will emerge as major international forces. However, the fact that several are being modelled directly on US investment banking lines suggests that their American counterparts have already established a commanding lead.

The Chief Executive of County Bank, Charles Villiers, made no apologies when he admitted that 'US investment banks are our model.' Nevertheless, the overall attitude towards US banks moving directly into foreign government bond markets appears to emphasise caution. In an interview in early 1986 the Chairman of Kleinwort Benson, Michael Hawkes, was quoted as saying 'We won't try to take them on in their own market at their own level, there's no point in us having a head-to-head with the mighty of this world, with Merrill Lynch or Goldman Sachs or Salomon. We're going to have to take them on here in London. We think that is enough.' Michael Hawkes' views may echo those of many other UK bankers, but Kleinwort Benson has already been successful in establishing niche participations in the highly competitive US market. It has one of the best swaps teams in the industry through its Los Angeles-based Kleinwort Benson Cross Financing and a thriving specialist futures subsidiary, Virginia Trading. Kleinwort Benson was also one of the first foreign banks to purchase a US government bond dealer, ACT1 Government Securities, which is now performing satisfactorily after being indirectly involved with two small US government bond dealers which failed in 1985.

If the identification of potentially profitable niche operations will be the selected strategy of most houses as deregulation becomes more widespread, the Euromarkets have provided numerous examples of how institutions can successfully develop new markets where none existed before. A Geneva-based bank, Gutzwiller, Kurz, Bungener, issued the first Eurobond denominated in Euro New Zealand dollars. Orion Bank, a London-based consortium now controlled by the Royal Bank of Canada, opened the Euro Australian dollar bond market. US Treasury warrants and stripped UK gilts were pioneered in London. The first note issuance facility was developed by Citicorp in London for a New Zealand government agency. So many houses claim to have invented swaps that it would be almost impossible to give credit to any one institution. The real origins of swaps can probably be found in the development of parallel loans in the 1960s and 1970s, a whole generation before swaps became a major industry.

The real lesson from the development of a market such as swaps is that imitation can be as successful as innovation. In the Euromarkets there is no such thing as concept copyright. A good new product can be directly copied, refined or repackaged within 24 hours. It must have been galling for a company such as Quadrex Securities Ltd, who can genuinely lay claim to having invented US Treasury warrants and stripped gilts, to have had to withdraw both new issues only to see them copied and successfully sold by another house a few days later.

Some institutions have cleverly combined swaps with an emerging new capital market to create their own successful and profitable niches. An example would be the Canadian Imperial Bank of Commerce which, through its London-based subsidiary CIBC Ltd, has become a significant factor in the once fast-expanding Euro Australian dollar bond market. Of even more interest is that a large number of CIBC Ltd's clients are Italian banks. To outsiders this

Rupert Hambro was certain that it would be difficult to compete against foreign institutions

would appear a highly unusual combination. Why would a Canadian commercial bank, not noted for any particular aggression in overseas markets, be arranging loans for Italian banks in a currency for which they almost certainly have no direct requirement? Enter the swap which is now the most important single influence in the Eurobond primary market. CIBC Ltd obtains the mandate from the Italian bank and immediately finds a counterparty or counterparties, probably based in Australasia, to assure the repayment of principal and interest in Australian dollars. In return the counterparty provides the Italian bank with alternative currency funds, probably floating rate US dollars at a level well below Libor. CIBC Ltd distributes the bonds through its own client base and through an international syndicate. CIBC Ltd also undertakes to provide a reasonable secondary market in that issue. CIBC Ltd will make money out of the swap fees and from its position as lead manager, particularly if it can distribute a signficant number of bonds to its own clients. The Italian bank gains wider recognition among the international investment community. The counterparty or counter-parties receive fixed-rate Australian dollars at a cost below that which they would have to pay in the domestic market.

Securitisation

Another strategy which will be pursued by many London houses in a deregulated market will be securitisation. This has become almost as popular a catch-phrase as globalisation which sounds very effective in speeches but is, in reality, within the reach or capability of no more than a dozen firms throughout the world. Securitisation on the other hand is within the reach of most innovative houses and the London institutions have already started to learn the rules in a sector where some American houses, in particular, excel. Securitisation essentially implies the transformation of an illiquid asset into one which is actively traded. Syndicated loans, once the corner-stone of international banking, have largely been replaced by floating rate notes and note facilities. The successful but also controversial sale of Rockefeller Center Properties by Goldman Sachs and Shearson Lehman showed the international investment community that property need no longer be a dormant low-yielding asset. In the United Kingdom some actual buildings and building developments have already been securitised. More will follow. Of particular interest to the foreign as well as the domestic banks is the vast UK mortgage market. In the United States the issuance of mortgage-backed bonds has become a multibillion dollar industry and numerous financial institutions are busy working on schemes by which UK domestic mortgages can be re-packaged and used as collateral for debt securities.

Some banks have actually bought UK estate agencies, including Hambros Bank who purchased Bairstow Eves, and Lloyds Bank who owns the Black Horse chain. Others are looking at the possibility of some formal link with the larger agencies with the view that a high proportion of house sales would be accompanied by applications for a mortgage. However, some have pointed out that the estate agency business remains highly fragmented and that even Bairstow Eves and Black Horse combined would account for less than 8 per cent of the sector. The most successful company in the United States in 'securitising' mortgages has been Salomon Inc who have adopted the approach in the United Kingdom by directly offering home loans to the public. Its UK-based subsidiary, the Mortgage Corporation, has taken prime television advertising space and hopes to lend £500 million in its first year.

The potential of securitisation seems almost limitless. The once totally illiquid syndicated loan market has been effectively securitised by houses such as Shearson Lehman Brothers who even quote a price for South American and Latin American debt. Original participants in these problem loans therefore now have a chance to reduce their exposure, albeit at a substantial discount from the purchase price.

In London the early trend of securitisation is likely to be towards mortgages or actual buildings. Eventually, however, this will progress towards other financial receivables such as car loans or hire-purchase agreements. It is conceivable that securitisations successfully applied in the broadest context could emerge as one of the fastest-growing sectors of the financial services industry.

Commercial paper

Having seen the extraordinary growth of the note facility sector of the Euromarket, it was almost inevitable that a market would emerge for sterling commercial paper. Taking into account all the attention which this has received in the press, it is worth reminding its many supporters that Eurodollar commercial paper was introduced by White, Weld & Co (now part of Credit Suisse First

It is conceivable that securitisation could emerge as one of the fastest-growing sectors of the financial services industry

Boston) in 1966. London is therefore only following an example set by the Euromarket 20 years ago. In the United States commercial paper was being issued in the early nineteenth century. The opening of a Eurosterling commercial paper market will be welcomed by both issuers and lenders. Prime credit issuers will be able to obtain cheaper funding from the banks and will presumably close some of their existing established credit lines. Institutional lenders will be able to achieve higher returns than on alternative short-term money market instruments or bank deposits. With the minimum investment likely to be in the region of £500,000, this is not a market which was designed for or which will appeal to the retail investor.

In the United States the present size of the commercial paper market may be approaching US$300 billion. It is therefore not difficult to see why UK bankers are so enthusiastic about the prospects for a similar opportunity in London, where most of the issuers of commercial paper will be their own long-standing corporate customers. No-one expects miracles but optimists talk of a £3–4 billion market within two years, and even these estimates may prove to be low in the same way that the size of the Euronote facility was grossly underestimated.

Some 30 UK companies or agencies already issue commercial paper in the United States and would presumably be automatic candidates for Eurosterling commercial paper programmes. As some standardised rating system will be required if this market is to be successful, those corporations and agencies which have issued commercial paper in the United States will have an advantage. The majority have already received prime credit commercial paper ratings in the United States from Standard & Poor's and Moody's. Less well-known UK borrowers might be reluctant to accept the rigidity of the US rating system and there could be an opportunity for a domestic rating company to classify some of the smaller issuers.

Although this is one of the few sectors in which the UK banks should be able to dominate after deregulation, there will still be competition from foreign institutions. Once again the US investment banking houses will be the main threat with their enormous experience in arranging and distributing commercial paper programmes. Some have already been granted official dealer status and if the market expands faster than expected and profit margins are acceptable, all the five main US dealers will join the new sterling commercial paper sector.

The Securities and Investments Board

The present reliance of the Euromarket on London will continue unless there are major negative changes in legislation. The role of the new Securities and Investments Board is therefore crucial, although, when the Securities and Investments Board (SIB) was

Sir Kenneth Berrill, Chairman of the Securities and Investments Board: 'You can't possibly do our job without criticism'

first announced, outside observers were alarmed by the absence of any recognised Euro practitioner. The Chairman of the SIB, Sir Kenneth Berrill, has a formidable academic background and was the senior partner of the London stockbroking house Vickers da Costa & Co, which was subsequently purchased by Citicorp. Despite being an early participant in the Euro sector, Vickers da Costa's strengths lay more in Japan, other Far Eastern securities and the London market. In an interview Sir Kenneth said that the expertise in Eurosecurities would originally have been supplied by John Abel, the former Chairman of Orion Royal Bank, who had subsequently resigned and had returned to Canada. Independent observers recall John Abel as an all-round generalist banker rather than as a specialised Euro practitioner.

It was inevitable that in its formulative stages the SIB attracted some considerable criticism. Comments ranged from 'a shambles' to 'totally ill-conceived' to 'the largest chicken and egg situation international markets have ever seen'. However, there was genuine sympathy for the need for regulation and for the difficulties facing the SIB. Sir Kenneth's own reputation within the Euromarkets has improved significantly following his open support for ISRO, the Euromarkets' own self-regulatory organisation (SRO). In addition, Sir Kenneth has appointed Archibald Cox, the senior Managing Director of Morgan Stanley International in London, to the board of the SIB. Archibald Cox is widely recognised as one of the most experienced Euro practitioners.

The Euromarkets cannot now complain that they are without representation on the Securities and Investments Board, and for their own SRO, the International Securities Regulatory Organisation (ISRO), which they expect to operate with a considerable degree of autonomy, they have been able to secure the services of Jonathan Agnew as Chief Executive. Jonathan Agnew is himself a recognised Eurobond authority, having been, like Archibald Cox, a Managing Director of Morgan Stanley International in London.

Sir Kenneth Berrill himself accepts the difficulties which confront the SIB. 'You can't possibly do our job without criticism. The main criticism will be when things start to go wrong.' In an interview he did not deny that the SIB would gradually assume a role similar to that of the SEC in New York. He was also in no doubt as to the extent of his authority. 'The Securities and Investments Board rules will take precedence – our rule is the one that you have to follow.' He was very concerned about the protection of retail investors in the Euromarket, even if they formed only a very small proportion of the buyers. Sir Kenneth conceded that some compromise was needed to differentiate officially between professionals and the 'Aunt Agatha'-type retail investors, and that 'professionals' would be defined in the Financial Services Bill.

The objections of the Eurobond primary market participants are that they cannot hope to maintain the market's efficiency as one of the world's best fund raising mechanisms if they are required to provide new issue information which would satisfy an SEC. Most Eurobonds today are swap driven and may therefore be announced at any time depending on market conditions being suitable for completion of the swap. The 'information gap' in the Eurobond market is notorious and needs substantial improvements. Lead managers should be required to provide information telexes on request and prospectuses or similar documentation after a period.

It should be possible, with only slight concessions on either side, to reach an agreement which will keep the Euro primary market operating smoothly out of its natural base in London rather than reiterating the previous threats to move the entire primary sector offshore to Luxembourg. The City rejected the Eurobond secondary sector once before and it would be foolish to do the same with the even more visible primary area.

Jonathan Agnew, Chief Executive of ISRO, is a recognised authority on Eurobonds

CHAPTER FOUR

The sterling capital markets

by Jonathan Orders*

This chapter concentrates on the 'Bulldog' market (domestic sterling market for foreign borrowers), the Eurosterling market (both fixed and floating rate) and concludes with a section on the sterling commercial paper market, which finally opened during 1986 amidst much public interest. It does not deal with the domestic sterling market for UK borrowers, often referred to as the debenture market (despite the fact that not all issues have been made on a secured basis). The approach of Big Bang has not, as yet, been felt as keenly in this market, whereas it has already made an impact both on the Bulldog market, where negotiated commissions and dual capacity already operate, and where methods of offering are currently in a process of change, and indirectly on the Eurosterling market, as evidenced by the growing number of foreign banks in book-running positions. Although not as many UK industrial companies as the authorities would have wished, issued debt in this market, during first half 1986 about £900 million had been raised, at maturities of five to 39 years (although most has been of more than 20 years' maturity) in principal amounts varying between £5 million and £100 million, and in both debenture and unsecured form. However, the majority of issuers have been property companies and investment trusts, rather than the large UK industrial companies. The UK debenture market is similar in most respects to the Bulldog market – the principal investors are insurance companies and pension funds, issues are typically placed on a yield basis, are listed in London (registered) and are frequently offered in partly-paid form. Many of them have, however, been placed by stockbroking firms alone, without the help of any UK bank (or syndicate of banks).

The role of the Bank of England

Access to the sterling capital markets, whether domestic or Euro, fixed or floating rate, is controlled by the Bank of England. Following the lifting in 1979 of exchange controls – which had been in place for 40 years – the Bank of England issued a series of Notices establishing the procedure for issuers intending to borrow sterling in either the domestic or Eurosterling market. The Bank of England Notice of November 1980, for example, reconfirmed that all new issues larger than £3 million would be subject to timing consent from the Bank of England under the Control of Borrowing Order 1958 (as subsequently amended). A New Issues queue was to operate for issues in sterling (apart from sterling Eurobonds) and in administering this queue the Bank would 'have regard to the capacity of the market and to the need to maintain orderly arrangements for new issues'. Eurosterling issues, however, would only require timing consent, as they were deemed not to affect the domestic market and therefore would not compete with the government's own borrowing programme.

A second important point made by this Notice referred to foreign lead managers: a UK-based institution must be lead manager, although a foreign bank was permitted to co-lead if a reciprocal arrangement was available in its own domestic market; otherwise a foreign bank could only be a co-manager. Finally, the Notice also extended the availability of gilt-edged status to issues of more than £7 million nominal (subsequently increased to £12 million) offered to the public by any overseas central government, whereas previously it had been confined largely to supranational institutions and Commonwealth governments.

These three principles have generally been maintained with occasional modifications, and have shaped the subsequent development of the sterling markets. Such modifications included a July 1982 Notice permitting companies to issue deep discount (including zero coupon) stocks, in order 'to facilitate stock issues by companies, as an alternative source of finance to bank borrowing'. This reflected a desire on the part of the authorities to see companies financing themselves longer term (the government's abstention from the long end of the gilt market for most of 1982–85 was similarly a response to this desire). The fact that permission to issue deep discount and zero coupon stocks was given to companies alone, and specifically withheld from local authorities – and therefore from other sovereign, parastatal or public sector borrowers – further underlines the authorities' stance.

Although all issues in the domestic market have been subject to the Bank of England's queue, it has normally been easier for private sector borrowers to arrange an issue date at short notice. Sovereign and supranational issuers may have to wait up to one year for a date. About six weeks beforehand, a firm new issue date is allocated, which will be the day on which the issue must be launched. There is, therefore, no guarantee that on the appointed issue date market conditions will be favourable, or that interest rates will be low. If a borrower decides not to issue on the date offered, he will usually be offered another date promptly if the reason for rejection is for bad market conditions alone. However, if

*Jonathan Orders is a Manager of Barings plc, Banking and Capital Markets Department.

the borrower declines to issue for any other reason – for example, the all-in cost may be considered too high despite market stability – then the authorities are unlikely to be as accommodating, although much would depend on how long the queue was at any one time.

In the Eurosterling market, issues can usually be timed at a borrower's discretion. Proposed issues are subject to formal timing consent from the government broker, under supervision from the Bank of England, but are not required to join the Bank's queue for issues in the domestic market. In practice, timing consent is readily forthcoming for Eurosterling issuers, with the Bank of England's main concern being to avoid issues being made simultaneously (leading to disorderly market conditions) rather than trying to ration access to the market. This is an important difference between the two markets, particularly now that a market for long-dated Eurosterling issues has been established. Flexibility of timing might give a borrower an opportunity to issue long-term debt at short notice instead of being forced to undergo the delays and uncertainties of joining the domestic queue.

A further constraint on access to the market by foreign sovereign and supranational borrowers exercised by the authorities concerns the frequency with which a borrower may approach the market, and the maximum principal amount which may be raised. Sovereign and supranational borrowers are restricted from coming to the Bulldog market more than once a year, for a maximum £100 million, and the Bank of England typically allows no more than one such issue per month. In the Eurosterling market access is less restricted, although the maximum principal amount per issue is again £100 million for sovereign and supranational borrowers. Moreover, the latter are prevented from using the sterling markets in general (i.e., Bulldog or Eurosterling) more than once every three months, so four sterling issues per annum raising a total of £400 million is the maximum amount that can be raised by a sovereign or supranational borrower. Corporate issuers, however, have considerably more flexibility.

The Bulldog market

The Bulldog market was reopened in August 1980, when the Kingdom of Denmark placed £75 million over 25 years through a small group of UK domestic banks in conjunction with a London stockbroker. The abolition of exchange controls by the first Thatcher government on its election in 1979 allowed the renaissance of the market. These controls had been in force for 40 years.

Exchange controls had prevented a foreign borrower (apart from those from the Scheduled Territories) from raising funds in the UK domestic market – except with specific permission from the Bank of England under the Control of Borrowing Order 1958 – although most investment managers were familiar with the characteristics of the market through their large holdings of issues by the UK government (gilts). However, the relative unfamiliarity of the new issuer, and a fall in the market, led to a poor reception to Denmark's issue, which was not, as became customary, offered at a yield margin above that of a comparable gilt, but at a fixed price. However, the market has developed and matured since then and by 30 June 1986, a further 57 Bulldog issues had been made, raising a total principal amount of £3.9 billion.

The borrowers

The Bulldog market has been dominated since its inception by sovereign (including sovereign guaranteed) or supranational borrowers, and the six issuers which followed Denmark into the market between January and May 1981 all fell within this category. Of these, three were made at the then more prudent five-year maturity, at margins of ½ per cent to ¾ per cent over the comparable gilt, compared with margins of at least 1¼ per cent in the case of the longer-dated issues. However, in July 1981 Inco, the Canadian primary metal and alloy producer, became the first corporate borrower to use the Bulldog market, with a long-dated issue incorporating a novel principal repayment option in US dollars at a fixed rate of US$1.97 to £1. Since that time only six other corporate issuers have come to the market. They have tended to be US companies, with hedging needs arising from net investments or specific acquisition needs in the UK. These companies were attracted, as are the sovereign and supranational borrowers, by the long maturities available and the opportunity to diversify their funding sources away from the typical Euromarket investors. For until very recently, the concurrently emerging Eurosterling market had offered new issue opportunities at 5–10-year maturities only, while the Bulldog market provided opportunities well beyond that maturity range, rivalled only by the US domestic market.

Initially, however, borrowers considered the use of the Bulldog market for five-year maturities as well as maturities beyond 20

Following Big Bang there will inevitably be changes in the way in which Bulldog issues are offered

years. During 1981-83 nine issuers tapped the market at five years, all of which were sovereign or supranational, or sovereign guaranteed: Sweden, World Bank (twice), Finland, New Zealand, Crédit National (twice), Malaysia and Spain. All these issuers, apart from Crédit National, have gone on to use the long end of the market, but the fact that there has been no five-year issue since the one by Spain in December 1983 underlines the rapid development of the market since that date. The institutional investors, who had for years concentrated on the gilt-edged market and therefore on unquestioned credit, were suddenly faced from 1980 with a succession of new credits to assess. Therefore many borrowers accessed the five-year market initially, to allow their stock to establish a trading pattern and to ensure investors' familiarity with the credit, before borrowing at the longer end of the market.

An additional way of preparing the market was to hold investor presentations, organised by the lead manager, where the borrower would address the main institutional investors in London and Edinburgh informally, sometimes over lunch, so that a mutually beneficial exchange of views and information could take place. Now the investors are sophisticated enough to accept new credits directly at the long end of the market, which would have been inconceivable a few years earlier. Investor presentations, however, are still of considerable value; their use is not confined to less well-known credits – the World Bank addressed the London investment community in Spring 1985 – and one recent long Bulldog issue by a new borrower received an adverse reception when it was sprung on the institutions without prior presentations.

Methods of offering

The two traditional methods of offering which have been used in the Bulldog market are the public offer for sale and the placing. Until July 1986 Bulldog issues have for Stock Exchange trading purposes fallen into two quite distinct categories: 'gilt-edged' and 'debenture' (i.e., non-gilt) stocks. Certain designated qualifying borrowers (which include sovereign states borrowing in their own names but not sovereign guaranteed agencies such as the French state borrowers) have preferred to employ the public offer for sale because under Stock Exchange rules they thereby became eligible to qualify for gilt-edged status, carrying the privilege of lower secondary market commissions (1/8 per cent against 3/8 per cent for an average £200,000 trade) and next day ('cash') settlement ensuring greater liquidity for the stock. Despite the higher advertising costs and primary market commissions of a public offer for sale (1¼ per cent against 0.66 per cent for a placing), eligible borrowers have opted for this method because the additional liquidity has allowed them to launch issues on rather lower margins above the reference gilt than would otherwise have been achieved.

The principal procedural difference between these two methods lies in the approach to investors: in a sub-underwritten public offer for sale, the initial approach to investors takes the form of an invitation to institutions from the managers, through the brokers to the issue, to sub-underwrite the issue (to the extent that the risk is not retained by the managers) for a generous commission of ¾ per cent, following which the managers seek applications for stock from investors, which may then be accepted or rejected. In a placing, the managers, via the brokers to the issue, make offers of fixed amounts of stock directly to investors at the issue price or margin, which, if accepted, form a binding contract – there is no sub-underwriting. The higher underwriting fees for a public offer for sale are justified on the basis that underwriters are at risk for a longer period (allowing time for advertising to the public) and are exposed to market as well as margin risk, because applications for stock are received only after final issue terms have been fixed. However, the attractive fees to sub-underwriters, most of which are also end-investors in the market (adding up to 0.1 per cent to the all-in cost), have ensured that few public offers for sale have encountered an unfavourable reception.

However, the established methods of offering Bulldog issues are currently changing rapidly. In May 1986 the Republic of Portugal made its debut in the market by issuing £50 million of partly-paid 30-year loan stock, by way of a placing. In many people's eyes this signalled the end of the offer for sale method. But for Big Bang, Portugal would most likely have made its debut issue by offer for sale. A further development took place in June when the European Investment Bank (EIB) approached the Bulldog market with a bought deal at a fixed price (the second time fixed pricing has been used since the original Denmark issue), as opposed to a fixed yield margin, under the lead management of a single UK merchant bank in conjunction with its wholly-owned broking firm. Commissions were the standard 0.66 per cent for a placing for a supranational borrower, but it is likely that commissions will be negotiated on new issues.

Some market commentators have suggested that all future Bulldog issues will be offered in this way, i.e., by fixed price placing

A queuing system is operated by the government broker in the interest of maintaining an orderly new issue market

underwritten and distributed by one bank. However, others feel that this is not the best way of achieving the widest possible distribution, which is in the interests of the borrower. Although this may be an appropriate method of offering and distribution for a well-known borrower it could be totally inappropriate for a lesser known, infrequent borrower. Fixed price issues may well become more common, since borrowers generally prefer to know the all-in cost of a borrowing at the outset, rather than be subject to market movements over a period of hours (or days, in the case of the old offer for sale). On Monday, 7 July 1986, in anticipation of Big Bang, the Stock Exchange permitted dual capacity and trading of Bulldog issues net of all expenses off the floor of the Stock Exchange, which has eliminated the principal advantage of the offer for sale (finer dealing commissions). The immediate effect of Little Bang, as it was called, was a six times increase in turnover in Bulldogs in the first week. The market has since settled down, however, with turnover up by about two to three times compared with the period prior to July. This is principally between market-makers, of which there are 12 in total, although most of the business is being done by about six of these.

Investors

Domestic UK institutional investors concentrate their purchases of fixed interest securities in two broad maturity bands: up to five and occasionally seven years, dominated by banks, building societies and the general funds of insurance companies, and 20 to 35 years. Between seven and 20 years domestic demand is much more limited, although some general insurance companies will take paper out to about 10 years when it offers a sufficiently attractive margin over gilts.

The long-dated securities are generally held as firm investments by a range of UK domestic investment institutions, primarily pension funds and life assurance companies, with an established pattern of long-term liabilities and a need to commit funds for periods of 20 years and more to obtain a certain yield to match those liabilities.

As mentioned above, the government has sought to create room for corporate issues in this maturity range by concentrating its own fund raising in the shorter maturities and in index-linked securities. Between late-1982 and end-1985 all but one of the conventional gilts issued with a maturity date beyond the year 2000 were small additional tranches of existing stocks which were swiftly absorbed

Breakdown of international sterling debt markets

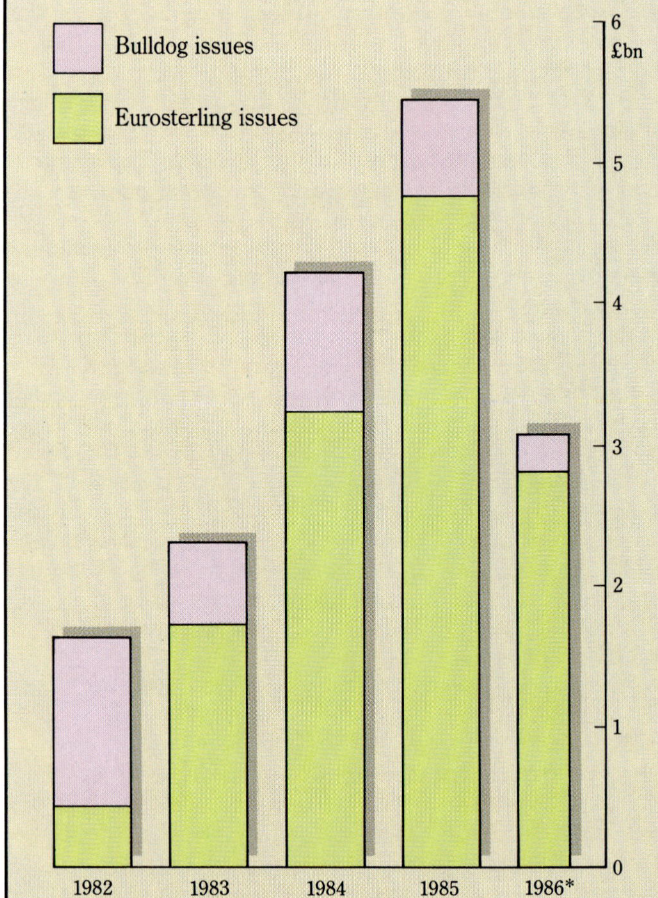

*Up to 15 July 1986

Source: Baring Brothers & Co., Limited.

Volume of bulldog issues

Breakdown by type of borrower

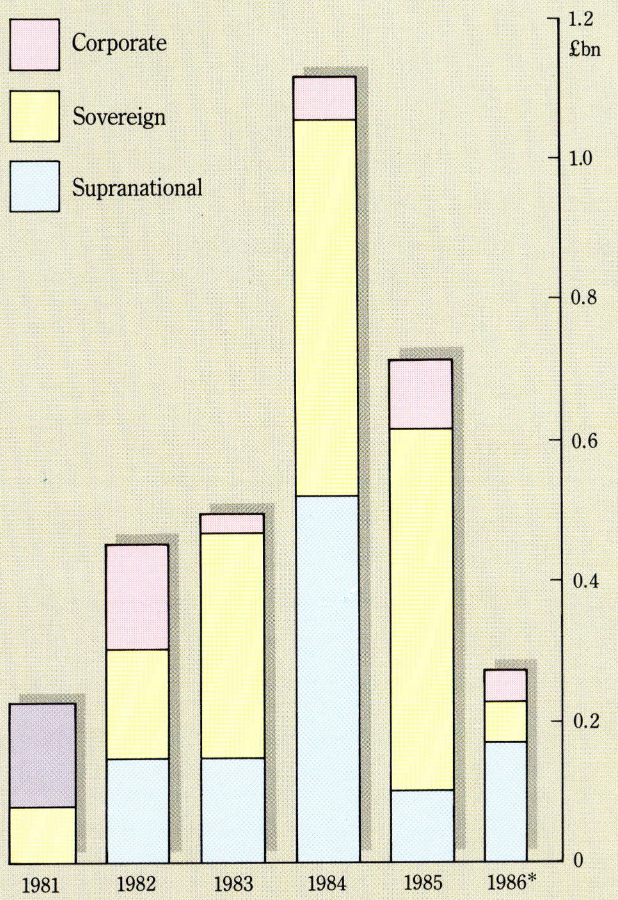

*Up to 8 July 1986

Source: Baring Brothers & Co., Limited.

by the market. There were thus opportunities for well-established domestic and foreign issuers to raise substantial amounts of sterling at these longer maturities.

Characteristics of Bulldog issues

Timing

All new sterling capital market issues for amounts of £3 million or more are subject to timing-consent from the Bank of England and, for new sovereign and supranational issues, a queueing system is operated by the government broker in the interest of maintaining an orderly new issue market. Where issues are placed in more than one tranche, separate timing consent is required for each tranche (and if a 'further issue' is made in the domestic market, it has to be queued in the same way as a new domestic issue). Under current practice, dates in the queue are allocated at the beginning of the month prior to the month in which the issue would occur, i.e. between four and eight weeks prior to the announcement date, although places may be obtained at shorter notice if another borrower holding a place withdraws.

Pricing

Since the first issue by Denmark, Bulldog issues, with the exception of the recent EIB issue mentioned above, have been priced 'on a yield basis' – i.e., at a fixed margin above the yield on a comparable gilt, typically the Treasury 13½ per cent 2004–08, although 'baskets' of stocks were also occasionally used up until the end of 1982. The coupon, payable semi-annually, and/or issue price are subsequently calculated on the basis of the issue yield and maturity date. New issue margins have narrowed considerably as the market has developed, reflecting in part the shortage of good quality long-dated fixed rate sterling paper, in view of the government's reluctance to use the market itself. To cite an example, in March 1986 the Inter-American Development Bank issued long-dated stock at a margin of 35 basis points over the reference gilt, whereas its previous two issues had been made at margins of 100 basis points (December 1984) and 135 basis points (November 1982). While changes in tax law were responsible for part of this decrease it nevertheless indicates how quickly the market has matured, particularly since early 1985.

Discount pricing has been prevalent in the Bulldog market, where sovereign and supranational borrowers are permitted to issue stock at a maximum discount of ½ per cent for each year of the issue's life, to a maximum of 15 per cent (i.e., minimum issue price 85 per cent). Domestic investors have been allowed capital gains on the discount totally tax free on such issues, and some institutions in any case prefer to buy stocks below par for accounting reasons. Consequently the vast majority of issues have utilised some sort of shallow discount. The tax treatment of both shallow and deep discount is described in detail in Appendix Two. Although zero coupon issues have not proved to be particularly attractive in the Eurosterling market, deep discount long-dated issues in the domestic market have been popular with investors since the issue by Safeway opened this market in October 1985.

Part-payment

By contrast with the Eurosterling market, where few issues have been launched in partly-paid form, long-dated Bulldog issues are made in partly-paid form (following the practice in the gilt-edged market), with 20–30 per cent usually payable initially and the balance three to six months later. The only exception to this rule has been the issue by Safeway in 1985, which was a deep discount issue. Part-payment is advantageous to investors when the yield curve is negative as it provides leverage which some investors find attractive and suits the cash flow of others. Bulldog issues are always issued in registered form, frequently with a bearer option to attract international interest.

Purchase funds

Bulldog issues made by offer for sale have frequently included purchase funds, which generally provide that during the first few years of an issue's life the purchase agent (invariably the lead manager) is authorised by the issuer to purchase and cancel a specific percentage of the issue if the price falls below the issue price. The purchase agent can thereby support the secondary market price of an issue, and investors are assured of some secondary market demand if the price falls below issue price. A purchase fund therefore helps to underpin the secondary market performance of an issue, and is therefore valued by investors. With the growing maturity of the market and the increased market making capacity resulting from Big Bang, it is unlikely that purchase funds will be used as often in future as they have been in the past.

Early redemption

Since long-term investors in sterling securities are chiefly institutional investors seeking to cover known and very long-term future liabilities, they are not generally prepared to accept the call options more common in the Euromarkets. UK institutions place a high value on having a certain repayment date, which also facilitates yield calculation and thus switching operations (which accounts for much of the liquidity in the Bulldog market). For the same reasons, put options have also been rare. As a result, nearly all Bulldog issues have 'bullet' maturity dates, although two French government guaranteed issues have included 'serial repayments' – i.e., repayment by four equal annual instalments during each of the last four years of the issue's life. Serial repayment is generally acceptable to investors since, unlike a sinking fund or early call feature, the dates and exact amounts of the repayments are known from the date of issue. However, early redemption for tax reasons is generally permitted in appropriate cases.

Size of issue and further issues

As mentioned above, all issues in the sterling capital markets by sovereign and supranational borrowers are subject to a maximum issue size limit of £100 million. It remains to be seen whether this limit – which has been in force for six years – will be raised in the near future. There is a possibility, too, that if a borrower is refinancing an issue which has matured, it may increase the principal amount of the new issue by that amount on the basis that these are not new funds being raised. For private sector borrowers, however, the £100 million limit does not apply, although it has been rare for them to raise more than £100 million in fixed rate form since their needs have rarely exceeded this figure. However, when BAT issued nearly £350 million of loan stock, raised for and secured by its acquisition of Eagle Star Insurance Co in 1984, the market capacity was available.

The ability of an issuer to place an issue of £100 million will clearly depend upon a number of factors, including market conditions at the time of issue, the attractiveness of the margin over gilts offered on the stock, and the perceived creditworthiness of the issuer. From the borrower's point of view, the economies of scale involved in the fixed costs of issue, as well as the improved liquidity of a larger issue, argue against small issues. The need for smaller amounts of long-term fixed-rate financing can sometimes be met economically through a private placement, but amounts as small as £10 million will often justify listing the issue.

Documentation now typically includes the provision for further issues of the same stock. This allows further tranches of the same stock to be issued in the future, at lower costs than would otherwise be possible. This technique has been tested twice in the Bulldog market: in August 1984 the World Bank made a further £100 million issue of its existing £100 million 11½ per cent Loan Stock 2003, creating the largest Bulldog issue (£200 million) in the market. More recently, in March 1986, the Inter-American Development Bank offered a further £75 million of its existing £100 million 9¾ per cent Loan Stock 2015. Both these further issues were well received by the market.

For issuers, the major advantages of the Eurosterling market have been its flexibility as to timing...

The Eurosterling market

The domestic sterling market was effectively closed to foreign issuers from the early 1960s until UK exchange controls were lifted in 1979. At the same time, international investors' generally negative views on sterling, and limitations placed by the Bank of England on sovereign and supranational borrowers, meant that the sterling sector of the Eurobond market was relatively slow to develop. The first Eurosterling issue, a 15-year £10 million issue by Amoco with a coupon of 8 per cent, was made in 1972, but activity lapsed after that year until 1977, when a little over £100 million was raised by way of public Eurosterling issues. Since then the market has been a regular source of funds for foreign borrowers and also for some UK corporate issuers.

The Eurosterling market bears a close similarity in its organisational and institutional aspects to the Eurodollar market. The Eurosterling market is also related to the domestic sterling market. Indeed, the abolition of UK exchange controls in 1979 removed many of the distinctions between the Eurosterling and domestic markets, and the same investors are now tapped by issuers in both markets, although a greater proportion of Eurosterling issues tends to be taken by offshore holders than is the case for domestic issues.

For issuers, the major advantages of the Eurosterling market compared to the domestic market have been its flexibility as to timing, its more relaxed attitude to security and the fact that under favourable market conditions funds can be raised more cheaply there than in the domestic market.

The fixed rate market

Method of distribution

Eurosterling fixed rate issues are arranged through traditional Euromarket management groups comprising around 10 to 20 banks (depending on the type of issue and market conditions) constituted under a lead manager. An issue might be sub-underwritten as to, say, 25 per cent to 35 per cent of the nominal amount by an international underwriting syndicate, and distributed through an international selling group. Sub-underwriting arrangements have been more prevalent in the sterling than in the dollar sector of the Euromarket, reflecting the market's greater orientation towards retail investors rather than the professional market. However, as the market has developed in recent years, issues have more commonly been fully underwritten by the management group, without a separate selling group.

There are three major areas of demand for fixed rate Eurosterling securities:

1 Offshore retail buyers, consisting of both expatriate British funds (invested, for example, from the Channel Islands) and foreign purchasers, particularly from Switzerland and the Benelux countries. These buyers are attracted by the traditional features of the Euromarkets, such as the availability of bearer paper, the payment of interest free of withholding tax and seven-day settlement. The 'foreign' element of this demand tends to be the more volatile, being influenced by expectations about the sterling exchange rate.

2 Offshore institutional investors, comprising principally insurance companies in Europe, Japan and the Caribbean. Some banks are also takers of Eurosterling paper at shorter maturities. These institutions are accustomed to dealing in the Euromarkets and are attracted by the same features that appeal to the retail market. Demand from this quarter is also sensitive to views about sterling.

3 British domestic investment institutions, notably the insurance companies and pension funds and, at the shorter end of the maturity range, the banks. These investors are very price sensitive and typically compare Eurosterling yields with those available in the domestic market, using gilt-edged yields as a benchmark; but as their sophistication increases, they are becoming more accustomed to trading on spreads rather than absolute yields.

Eurosterling (particularly corporate) issues tend to be placed predominantly among the first two groups, comprising the traditional Euromarket buyers, with the extent of domestic interest depending very much on yield relativities. An issue in the domestic sterling market, by contrast, needs to be priced to appeal to the domestic institutions and they usually play a more significant role in the market.

Amount

Recently fixed rate Eurosterling issues have ranged in size between £20 million for Privatbanken in March 1985 and several issues of £100 million for the World Bank and other commodity borrowers.

Given the size of the fixed rate Eurosterling market at present, it

The offshore element of the Eurosterling market is thinner than the domestic markets

would be impractical to expect it to absorb a much larger offering than £100 million. To date, offerings over £30 million have generally been followed by a liquid secondary market. However, it is possible, within these constraints, to raise considerably more than £100 million from this market through a series of issues.

Maturity

Until the March 1986 Budget, Eurosterling issues had maturities in the range of five to 10 years, with an occasional issue which had a 12-year maturity. However, during the recent past more flexibility has been evident with regard to issues at both the short and long end of the market, as described below:

One to five years By a statutory instrument which came into effect on 24 April 1985, regulatory changes were made to enable companies to issue sterling denominated bonds with maturities of between one and five years. The key points are:

- the market is only available to companies with equity listed in the United Kingdom or their guaranteed subsidiaries;
- both fixed-rate and floating-rate bonds may be issued;
- a prospectus (or listing particulars) must be prepared;
- to allow interest to be paid gross, bonds must be listed in London or Luxembourg;
- bonds may be issued either by an underwritten issue, or on a tap basis over a period of time;
- there are no limits on issue size;
- bonds must be issued in minimum denominations of £100,000, at a minimum maturity of one year; and
- timing consent must be obtained from the Bank of England, but a blanket authorisation for a period may be obtained.

The government's aim with this legislation was to allow companies to reduce their borrowing costs by borrowing directly from non-bank investors. However, the market remains unused to date, as companies have hitherto been able to finance themselves more cheaply through banks and in the sterling swap market, without having to undergo the cumbersome listing and prospectus requirements.

10 to 20 years Prior to the March 1986 Budget, the Eurosterling market had not reached much beyond 10 or 12 years, with the Bulldog market being used for maturities of 20 years and beyond. However, the March 1986 Budget introduced a ½ per cent stamp duty on transfers of registered loan stock over five years' maturity. This adversely affected the domestic market, where all issues are in registered form. As an immediate result, ICI issued £100 million 17-year bonds in the Eurosterling market, precisely to avoid this punitive stamp duty. The issue was made without the supporting covenants which UK domestic issuers usually provide when issuing long-term paper in the domestic market, but was well received. Since March 1986 there have been a further seven long-dated issues, all by UK companies, none exceeding maturities of 20 years. Only the issue by Wellcome was able to match ICI's reception. Some of the other issues immediately traded at significant discounts to their issue price, as the UK domestic institutions – the main investors in this long-term debt – reacted against the new trend.

The authorities have subsequently reinstated the stamp duty exemption, but the maturity limits of the Eurosterling market have been irrevocably extended. The Eurosterling market is now a five- to 20-year market, and although no foreign borrower has issued at the longer maturities, they are now able to choose between the Bulldog and Eurosterling market when issuing long-term sterling debt.

The optional redemption provisions available in other Eurobond markets can be negotiated for Eurosterling issues from five years, if desired. However, early redemption options are in general unpopular among investors in the sterling capital markets (particularly among the domestic institutions). The inclusion of these features, while feasible, in most cases reduces market appeal.

Timing

Eurosterling issues can usually be timed at a borrower's discretion. Proposed issues are subject to a formal timing consent from the government broker, under supervision from the Bank of England, but they are not required to join the Bank's queue for issues in the domestic market. In practice, timing consent is readily forthcoming for Eurosterling borrowers. The Bank's concern is simply to avoid clashes between issues that might lead to disorderly market conditions, rather than trying to ration access to the market.

The preparation of documentation for an initial issue by a borrower generally takes up to two weeks. If documentation has been prepared in advance, a Eurosterling issue can be brought to the market immediately, assuming timing consent is forthcoming.

Eurosterling fixed interest issue volume

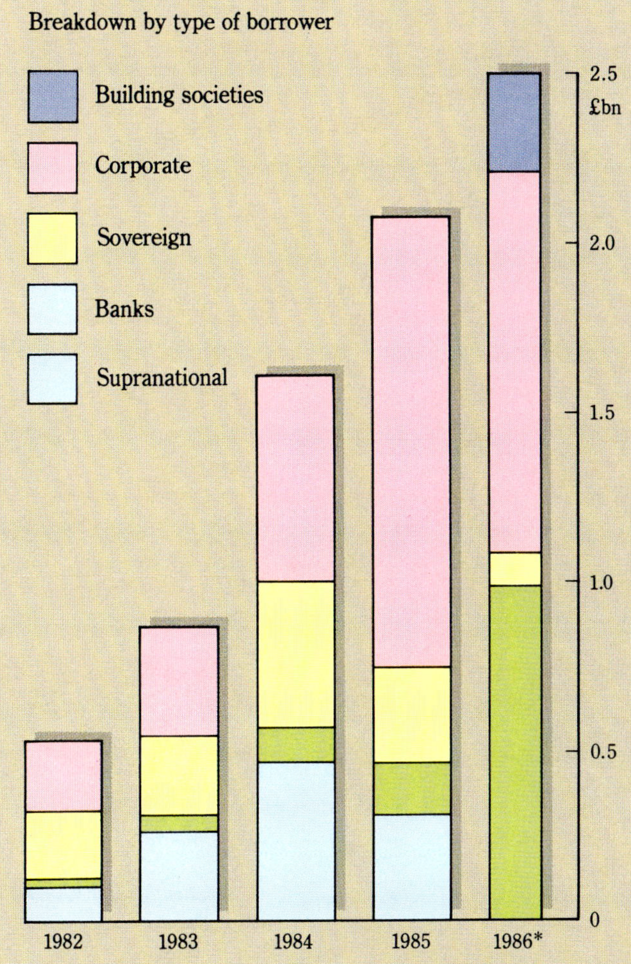

*Up to 8 July 1986

Source: Baring Brothers & Co., Limited.

Distribution of an issue generally takes a week, and payment is made to the issuer within three to four weeks of the date of launching the issue, although the settlement date can be fixed largely at the discretion of the issuer and lead manager.

Pricing

Issues in the Eurosterling market are typically made on a fixed price basis as is usual in the Euromarket. A benchmark used in pricing Eurosterling issues is the yield available on gilts of comparable maturity. However, there are sufficient differences between Euro and domestic securities, and between the investors in these, to permit the Eurosterling yield curve to differ from the gilt yield curve, and for some prime-quality Eurosterling issues often to trade at yields below those on corresponding gilts. These differences reflect, among other things, views taken by international investors on the likely future course of sterling and the absence of UK withholding tax on Eurosterling securities.

The offshore element of the Eurosterling market is, however, thinner than the domestic markets, and demand may be somewhat volatile, depending in part on currency exposure decisions by international fund managers. For that reason, the most favourable pricing in the Eurosterling market is generally available to those borrowers who are able to move quickly when a suitable opportunity arises. Yields on secondary market issues are obviously also carefully scrutinised when pricing a new issue, although the yields on seasoned issues are frequently misleading as some issues are more firmly placed than others, which clearly affects their individual trading performance without having any significance for a new issue.

Deep discount pricing

Changes announced in the 1984 Budget and passed into legislation in the 1984 Finance Act cleared the way for deep discount and zero coupon issues to be launched in the sterling markets by corporate borrowers and banks. The first Eurosterling zero coupon issue was made by Redland Finance plc in March 1985. Sovereign and supranational borrowers, however, are still prevented by the Bank of England from making these issues.

Briefly, 'shallow discount' issues, as defined in the Act, are those that have an issue price that represents a discount from par of no more than ½ per cent per year to maturity (subject to a minimum issue price of 85 per cent). A shallow discount is treated by the Inland Revenue as a capital item for tax purposes and is not deductible by the issuer for UK tax purposes. Deep discount issues, i.e., those with an issue price lower than the limits mentioned above, have their initial issue discount treated as income for taxation purposes. The tax treatment of discount issues is summarised in Appendix Two at the end of this Chapter.

Security and covenants

In keeping with usual Euromarket practice, issues made or guaranteed by borrowers of high credit standing are without security, although typically with a negative pledge provision.

Had the City become a follower rather than a leader?

Form of issue

Securities are issued in the form of bearer bonds, constituted by a trust deed (or issued with the benefit of a fiscal agency agreement), and generally in denominations of £1,000, £5,000 or both. Interest is paid annually in arrears. Arrangements for part-payment by investors at time of issue with the balance payable later are common in the domestic sterling market, but that feature has not gained more than passing popularity in the Eurosterling market. However, there may be occasions when part-payment could offer some savings in borrowing costs.

Listing

Since the beginning of 1985, when the new Listing Requirements for the London Stock Exchange came into force, an increasing number of Eurosterling issues have been listed in Luxembourg. Where an issuer does not already have a London listing (which reduces the burden of listing requirements for a new issue), Luxembourg is often preferred to London.

Structure of commissions

	5 years	**7 years**	**10–12 years**	**17–20 years**
	%	%	%	%
Selling concession*	1¼	1¼	1⅜	1½
Management fee	¼	¼	⅜	½
Underwriting fee	⅜	⅜	⅜	½
Total	1⅞	1⅞	2⅛	2½

*Generally passed on to professional investors by the members of the selling group.

A sterling swap market has grown in parallel with the development of interest rate and currency swaps in the international markets

Commissions and costs

Eurosterling issues involve higher commissions than are charged in the domestic market but they are reallowable. Similar differences arise in the corresponding dollar markets. Reflecting the smaller issue volumes in the Eurosterling markets, however, commissions in the sterling sector are a little higher than in the dollar sector at some maturities.

Issue expenses are also similar to those arising with Eurodollar issues and they vary, of course, with the size, complexity and maturity of an issue and according to the status of the issuer. The World Bank, for example, has issued 5-year Eurosterling bonds at commissions of 1¾ per cent and 10-year bonds at commissions of 2 per cent.

The listing particulars, which provide information on the issuer and the security offered, may consist of an offering circular and an exchange telegraph (Extel) card or just an Extel card alone.

Extel cards are required by the Stock Exchange for all securities listed in London for distribution to professionals in the market. For each issuer there are usually two cards. One gives financial and other information on the issuer based on its most recent annual report and any subsequently available financial information. The second card provides details of the securities being offered and contains the listing particulars which are submitted to the Stock Exchange and circulated to market professionals.

Secondary market

Secondary market trading is carried out in an over-the-counter market, notwithstanding the listing of the securities on a stock exchange. Eurosterling securities may be held in either of the Eurobond clearing systems, Euroclear or Cedel. As mentioned above, Eurosterling issues are more attractive to investors if they hold out the prospect of reasonable secondary market liquidity.

Swapped issues

A sterling swap market has grown in parallel with the development of interest rate and currency swaps in the international markets. The market is principally based in London. It is smaller and less liquid than the dollar swap market, reflecting the relative sizes of the capital markets in the two currencies.

Rates for sterling swaps have in the past fluctuated very considerably in relation to yields on gilts. The source of fixed rate sterling in the swap market has frequently been swapped Eurosterling issues; typically, the lead manager of the issue will underwrite the swap rate and then seek to cover the swap in the market. The payers of fixed rate sterling in the swap market tend to be a wide range of financial and corporate names, looking to fix liabilities in smaller amounts than the typical size of an issue. Intermediaries in this market are, therefore, often faced with the need to take swap positions onto their own books in order to break them into smaller amounts. The correlation between swap rates and Eurosterling rates has tended to be closer than the correlation between either and gilt yields. This has, from time to time, provided opportunities for issuers in the Eurosterling market to raise floating rate funds via a swap at margins below sterling Libor. The opportunities usually arise when international demand for Eurosterling bonds is strong. At these times there can be a flood of issues, until either the demand for bonds has been met, or swap rates have fallen to remove the arbitrage opportunities.

In 1986 the principal borrowers of floating rate Eurosterling have been building societies. In March 1986 they were allowed to write interest rate swaps for the first time, and since then there have been five fixed rate issues for building societies which have been swapped to provide floating rate funds at or below Libor, offering good cost savings over the floating rate note market.

Equity-linked issues

These have been somewhat rare in the sterling markets. During 1980 and 1981, about 10 Japanese companies made equity-linked offerings, all in convertible form (equity warrant issues were prohibited for Japanese companies by their authorities) at 15- or 16-year maturities. Some of these were offered in the domestic, and some in the Eurosterling market, and they met with varying degrees of success. However, as Japanese securities houses were unable to lead manage any of these issues, which had (as now) to have a UK bank as book-runner, and as cheaper opportunities existed in the US dollar market, the sterling equity-linked market has been largely neglected by Japanese companies since 1981.

The equity-linked market was reopened in July 1983 when Schering brought a 7-year equity warrants issue to the Eurosterling market. Since then there have been issues, including in July 1985 a convertible issue by Pirelli, with simultaneous convertibles in US dollars and Deutschmarks and, earlier this year, an equity warrant issue by Yamaichi Securities, the first Eurosterling issue to have a Japanese house as book-runner.

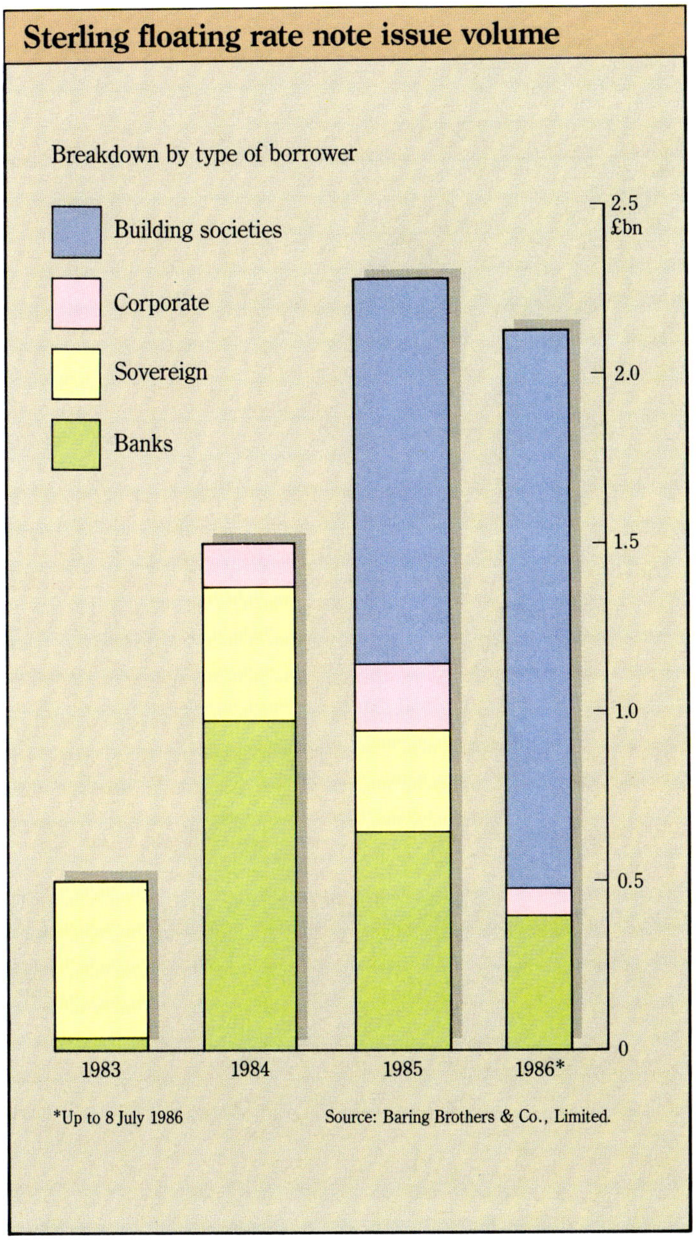

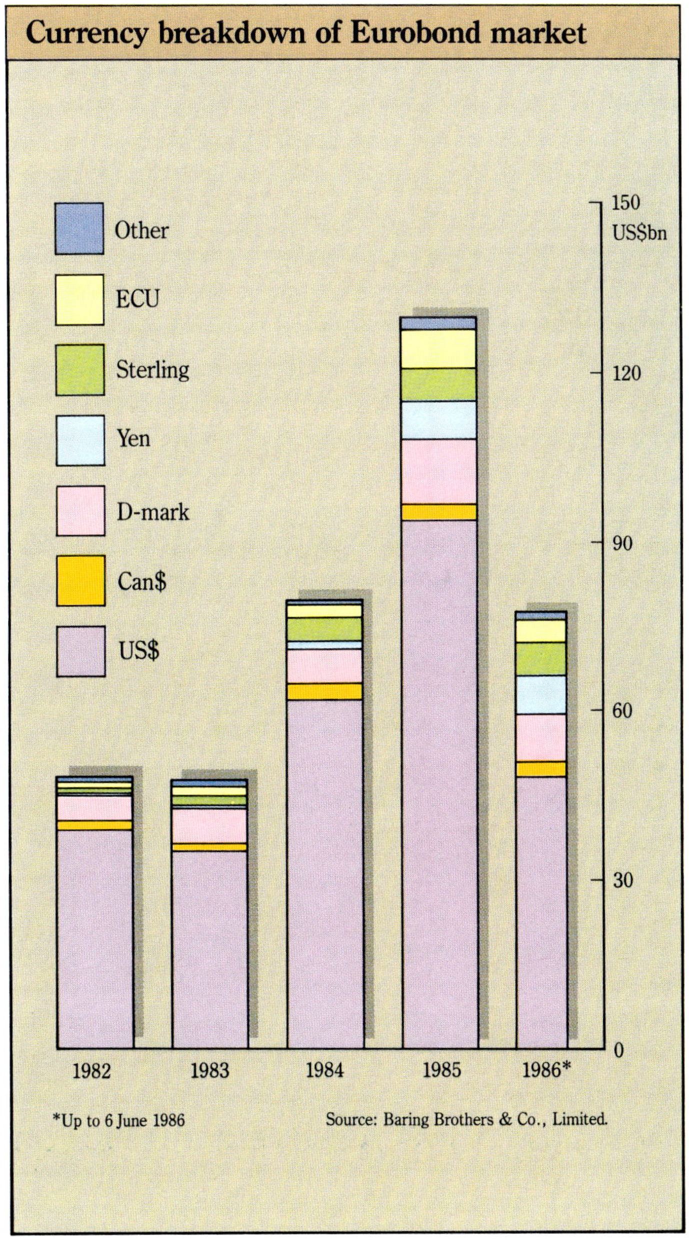

The floating rate market

The Eurosterling FRN market came into existence in mid-1980 when two banks, Scandinavian and Lloyds, raised £20 million and £50 million respectively. However, largely due to illiquidity, and the long maturity of these issues (both 10-year maturities, with no investor's put option), the sterling FRN market then closed for a number of years. It remained dormant until 1983, when a number of changes in the UK market created a much more propitious environment for the issue of floating rate sterling paper.

While there had been continuing demand for floating rate sterling assets within the London banking community, the supply had been constrained by (1) the withdrawal of local authorities from bank borrowings since they were now funding themselves from central government sources, (2) the relatively low borrowing activity of the UK corporate sector and (3) efforts by the Bank of England to slow the growth of the UK acceptances market. Demand for floating rate paper more generally had also been stimulated by the need to match asset/liability maturities along with the growth in balance sheets generally.

The first response to these factors was a successful syndicated bank loan for the Kingdom of Sweden in late August 1983; the initial offering was doubled to £500 million. That was quickly followed by the reopening of the sterling FRN market the following month, when Société Nationale des Chemins de Fer Français, guaranteed by the Republic of France, successfully issued £75 million. Since then there have been some 60 issues raising over £6 billion.

The earlier issues were easily placed, reflecting the demand which had built up between 1980 and 1983. The issues since have been launched into a more mature market, in which correctly priced issues from a broad spectrum of borrowers continue to attract strong investor interest.

As with the US dollar FRN market, there has been a marked decline in the all-in cost to the borrower as the sterling FRN market has developed and matured. In most respects the Eurosterling FRN market is similar to that for Eurodollar FRNs, as will be apparent from the summary below. However, an important difference between the two markets, which affects yields over Libor, is the requirement that bank holders of sterling FRNs subject to regulation by the Bank of England hold interest free balances with the Bank of England in proportion to their eligible liabilities, and will in the future have to hold designated assets for liquidity. The cost of similar requirements in the past has varied from time to time and has recently been about 1/16 per cent per annum. This cost can be expected to continue at a similar level in the future. Reserve asset costs are not incurred by regulated UK banks in respect of their holdings of dollar FRNs.

Since early 1985 there have been a number of significant developments. In January and February 1985 two successive 2 per cent rises in UK base rates kept issuers away from the market until a fall in base rates and a levelling out of the negative sterling yield curve brought a succession of banks to the market. Since Autumn 1985 there has been a steady flow of issues by UK building societies, following an Inland Revenue Statement that from April 1986 UK building societies would be permitted to make interest payments on Eurobond issues without deduction of tax at source. Since the announcement, 11 building societies have brought issues to the market, raising over £3 billion in 10 months. This demonstrated the market's capacity to absorb large quantities of paper from similar borrowers due to new investors – and with them new market makers – coming to the sterling market as yields on triple-A US$ FRNs have moved below Libid in many cases.

Amount

Sterling FRN issues have ranged in size between £30 million and £500 million, although sovereign and supranational borrowers are still limited to £100 million per issue. However, since the advent of the building societies, issue sizes have increased, culminating in the £500 million issue by Abbey National in August 1986. One way to overcome the £100 million limit on sovereign and supranational borrowers is to provide for a further issue of the same notes at a future date. This technique was used by New Zealand in November 1985, which thereby created a single issue of £200 million. Smaller issues of between £30 million and £50 million do not trade very actively in the secondary market.

Maturity and call provisions

Demand for sterling FRNs is concentrated in the five- to 12-year maturity range. While longer maturities have been achieved, the market has shown some resistance to longer-dated issues which do not incorporate investors' put options. For most borrowers, the minimum maturity remains five years, since the Bank of England does not permit issues of securities with final maturities (or put or call options) of less than five years unless the issuer complies with the burdensome requirements of the short term corporate bond market (see above) from which public sector borrowers are excluded. Call options on floating rate issues of longer than five years are fairly standard, in contrast to the fixed rate market. Only one perpetual or undated issue has been attempted in the sterling FRN market, when Standard Chartered issued £150 million (plus £150 million 'tap') perpetual in May 1985. The issue was placed only with great difficulty, and has been illiquid ever since.

Interest determination

The interest rate on sterling FRNs is usually set by reference to

Sterling commercial paper will have to be competitive with the sterling eligible bill market

domestic sterling interbank rates. Most of the issues so far have used the London interbank offered rate (Libor), although the mean of the offered and bid rates (Limean) has also been used. The usual spread between the bid and offered rates at three months is ⅛ per cent.

Recently the sterling reference rates used have usually been for three-month deposits, in contrast to the common, but by no means exclusive, practice in the Eurodollar FRN market of pricing off six-month Libor. This is a consequence of generally greater liquidity in three-month sterling time deposits.

As in the Eurodollar FRN market, investors expect to see a minimum interest rate incorporated in the terms of an issue – in most current issues these range between 5 per cent and 5½ per cent.

Security and covenants
As in the Eurosterling fixed rate market, cross default and negative pledge clauses are usually included in FRN documentation.

Listing and marketability
The majority of foreign borrowers who have tapped the sterling FRN market have opted for a listing on the Luxembourg rather than the London Stock Exchange, since, as mentioned above, disclosure requirements for a London listing are considerably more onerous and fees payable to the London Stock Exchange are marginally higher.

Commissions and costs
The total return to an investor from an FRN reflects both the interest paid and the difference between the purchase price (or price at the selling concession) and the redemption price, amortised over the life of the issue. Commissions vary in accordance with maturity, the market's perception of the issuer and how aggressively the interest rate margin is pitched. Since early 1985 commissions have diminished as issues have been successfully launched on increasingly tight terms. However, margins are still significantly above those in the Eurodollar FRN market, where a sovereign name can frequently issue paper at Libid, compared with a margin over Libor in sterling. Recent issues have been fully underwritten by a large group of co-managers alone, the size of the group (25-30) depending on the size of the issue.

Sterling commercial paper

In May 1986 the Bank of England finally published guidelines on the issue of sterling-denominated short-term debt securities (sterling commercial paper). This followed a period of increasing interest by companies and banks in the United Kingdom in the creation of a market in such securities. The guidelines, summarised below, restrict potential issuers of sterling commercial paper to major UK and overseas companies listed on the London Stock Exchange.

Market developments

Sterling commercial paper, if it is to be attractive to issuers, will have to be competitive with the sterling eligible bill market, which is the principal alternative source of floating rate funding for UK companies. The bill market is attractive because of the rates at which bills can be discounted. These rates reflect eligible status, which allows them to be discounted at the Bank of England. The Bank has been a major buyer of bills, and will continue to be so for some time even if the bill mountain diminishes. The Bank's presence as a major buyer has inevitably affected rates in the bill market, which from time to time have been well below Libor. The intermediary costs paid by top-quality borrowers in the bill market are very low, lower than dealer's commissions have historically been in the US commercial paper market. Sterling bills drawn by a company should always trade at levels equal to, or lower than, the same company's sterling commercial paper because of the security offered by the additional name on the bill.

In the future the eligible bill market may not continue to offer the same savings, relative to Libor, that it has in the past three years. Furthermore, the eligible bill market is most attractive in the one to three-month maturity range. Companies wishing to fix rates for longer periods very often borrow through one to three-month bills, and cover future periods with future rate agreements. Sterling commercial paper may be issued for periods of up to one year, and this may provide advantages.

However, the principal advantages are likely to be at the very short end of the maturity spectrum, where the use of the sterling commercial paper has already provided benefits to some borrowers. Many companies have monthly fluctuations in cash flow which require marginal borrowing for short periods within the

The longer-term development of the market will depend on the willingness of a number of major companies to approach this market

month. The bill market is not available for maturities of less than 14 days and often offers little advantage over money market borrowing for periods of less than one month. Within this period, therefore, the ability to issue sterling commercial paper may provide cost savings.

In comparing sterling commercial paper with the bill market, the principal cost savings come from disintermediation. There are of course many companies which are unable to use the eligible bill market, because they have no, or an insufficient number of, suitable transactions against which to draw bills. Such companies may well be the principal issuers of sterling commercial paper in its initial period.

The longer-term development of the market will depend on the willingness of a number of major companies to approach this market to provide benchmark issues, and a regular supply of paper to investors. According to Bank of England figures released in July 1986, sterling commercial paper outstanding on 18 June 1986 amounted to £50 million, £20 million of which was left with the banking system itself. There then appears to have been an acceleration during July, and independently produced figures suggested that £175–200 million of paper had been issued by mid-July (including a £60 million placement by Allied-Lyons), of which £100–150 million was still outstanding. Of the 21 programmes announced by this date, 18 were for UK groups.

APPENDIX ONE

Summary of CP regulations

Issuers

- UK or overseas companies listed on the London Stock Exchange and with a net worth of £50 million or more; and
- wholly owned subsidiaries of such companies provided that their obligations are guaranteed by the listed parent company.

NB: (i) Banks, building societies, etc. which can already issue CDs may not issue commercial paper. (ii) Issuers can arrange for their paper to be guaranteed by a bank but the issuer must still meet the above requirements.

Denominations

Paper must be issued and transferable in minimum amounts of £500,000.

Maturity

Paper must have a maturity of between seven days and one year.

Management of issues

- Paper can be issued direct by issuers to investors; or through an intermediary (which for the time being must be a UK incorporated bank or a licensed deposit taker currently regulated by the Bank of England).
- Intermediaries can act either as agents to place paper on behalf of the issuer or as principals with a commitment to buy or underwrite the paper.

Disclosure

- Until the passing of the Financial Services Bill, issuers in the United Kingdom will have to comply, where appropriate, with normal prospectus requirements for debt securities. To avoid these requirements sterling commercial paper has been issued orally without offer documentation.
- Overseas companies (including subsidiaries of UK companies) already have a dispensation from prospectus requirements under Section 79(2) of the Companies Act 1985, provided that the paper is offered only to 'securities professionals'.
- Where issues are made without a prospectus, a statement must be printed on the paper stating that Stock Exchange listing regulations have been met and no adverse change has occurred since date of last publication.

Tax

- Interest on sterling commercial paper can normally be paid without deduction of income tax.
- Interest/discount incurred by the issuer as a trading expense will normally be deductible in computing profits for corporation tax.
- The arrangements for taxation of the return earned by holders of sterling commercial paper are those applying to certificates of deposit.
- The issue and transfer of sterling commercial paper will be free of stamp duty.

APPENDIX TWO

Summary of tax treatment of sterling fixed interest securities in the UK

Chronological sequence of principal changes in legislation

Position prior to March 1984
All fixed interest securities subject to capital gains tax except gilts held for over one year. Gains on fixed interest securities held for more than one year eligible for indexation against Retail Price Index (RPI); could reduce a gain but not offset a loss.

Issue of deep discount securities by companies permitted but tax rules unclear.

March 1984 Budget
Capital gains tax exemption for gilts held for more than one year extended to 'qualifying corporate bonds' only.

Tax treatment of deep discount securities clarified.

28 February 1985
Announcement of proposed abolition of 'bondwashing', to take effect with introduction of the 'accrued income scheme' on 28 February 1986. 'Forestalling provisions' to apply until that date.

March 1985 Budget
Indexation extended to cover all disposals of fixed interest securities regardless of period held. Losses against RPI to be available to offset gains elsewhere.

2 July 1985
Announcement of proposed abolition, from 2 July 1986, of all capital gains tax on qualifying corporate bonds. Indexation no longer to apply to such bonds.

28 February 1986
Accrued income scheme becomes effective.

March 1986 Budget
Announcement of imposition, effective immediately, of ½ per cent stamp duty on transfer of registered loan stock except that issued by certain supranational borrowers, and 5 per cent tax imposed on conversion of shares into ADRs

29 April 1986
Announcement of repeal of above stamp duty. Tax on ADRs reduced to 1½ per cent.

2 July 1986
Abolition of capital gains tax on qualifying corporate bonds becomes effective.

Separation of income from capital gain

Tax treatment of issue discounts
Fixed interest sterling securities other than UK government gilts are classified as 'deep discount' bonds where the original discount to par exceeds 15 per cent of the redemption amount, or more than ½ per cent for each year of the life of the security.

Deep discount
In the case of a deep discount security, the original issue discount is treated as income rather than capital gain, but for investors (other than, in the view of the Inland Revenue, certain financial traders), income tax is only payable on the accrued discount when the security is redeemed or sold. For a tax-paying investor subject to a constant or declining income tax rate, this postponement of income tax liability gives a higher after-tax return than a current coupon bond issued on the same yield to maturity. Conversely, a UK issuer is permitted to make an annual deduction against tax.

Deemed interest is calculated according to the yield to maturity of the bond at issue and remains constant, regardless of the secondary market performance of the bond. Hence an investor who sells at a price below his acquisition cost is still liable for income tax on the deemed interest received. Any gains above the deemed interest fall under the capital gains tax provisions.

Shallow discount
Shallow discounts are treated as capital gains for tax purposes. UK issuers of shallow discount securities therefore obtain no tax deduction for the shallow discount element.

Bond washing
Bondwashing refers to the practice of converting accrued interest into capital gain for tax purposes. Until February 1986 bondwashing was prevalent in the United Kingdom, where tax-paying investors were able to benefit from the favourable tax treatment of capital gains by purchasing securities ex-dividend and selling one year and one day later cum-dividend.

However, from February 1986 accrued interest, on a day-by-day basis, has been treated as income as set out in the accrued income scheme. Until that date, interim forestalling provisions applied, which effectively prevented bondwashing during the intervening period.

Capital gains tax and indexation

In the 1984 Budget, capital gains tax exemption for fixed interest securities held for over one year was extended from gilts to qualifying corporate bonds. A security is classified as a corporate bond when it fulfills *all* of the following:

(a) it is issued by an entity any of whose debt or equity is listed or traded on the Stock Exchange;
(b) it represents a 'normal commercial loan' (i.e., it is in no way equity linked); and
(c) it is in sterling, with no provision for conversion into or redemption in another currency (other than redemption in a foreign currency at the then prevailing exchange rate).

A qualifying corporate bond is a corporate bond either issued or acquired by the holder after 13 March 1984.

At the time of the 1984 Budget, the shelter of capital gains by indexation against the Retail Price Index (RPI) was available only for gains made on bonds held for more than one year, and was not applicable to losses or to short-term gains. These limitations were due to be removed following the 1985 Budget, creating a very favourable tax regime for qualifying corporate bonds, where gains would be exempt from tax and losses against the RPI could be utilised to shelter gains made elsewhere in an investor's portfolio.

However, on 2 July 1985 the Inland Revenue moved to correct this situation by abolishing all capital gains tax on qualifying corporate bonds with effect from 2 July 1986 and by stipulating that indexation would not apply to them. This has resulted in quite separate tax treatment of qualifying corporate and non-qualifying corporate bonds, since the latter are not capital gains tax exempt, but may be indexed against the RPI in respect of both gains and losses over any period.

Deduction of interest at source

Domestic sterling bonds
In the domestic sterling bond market, interest payments are normally made net of 29 per cent withholding tax, which is the basic rate of UK income tax. This requirement does not apply to the following borrowers:

- International Bank for Reconstruction and Development;
- Inter-American Development Bank;
- Asian Development Bank;
- African Development Bank; and
- European Investment Bank.

In the 1985 Budget, the exemption was extended to include three European institutions:

- The European Economic Community;
- Euratom; and
- The European Coal and Steel Community.

Eurosterling bonds
In the 1984 Finance Act permission was granted to UK companies to make payments on quoted Eurobonds in bearer form without deduction of income tax at source where the paying agent is overseas or the beneficial owner is not resident in the United Kingdom or the bond is held in Euroclear or Cedel. The ability of non-UK issuers to pay interest on Eurosterling bonds free of UK withholding tax was not affected. This change has largely eliminated the need to use an offshore finance subsidiary, but there remain some circumstances in which an overseas issuing vehicle is advisable. The 1985 Budget announced that UK building societies (which were not included in the 1984 exemption and are not permitted to establish offshore finance subsidiaries) would be allowed to pay interest gross on bearer bonds from 6 April 1986 pursuant to the Income Tax (Building Societies) Regulations 1986. This opened the way for Eurobond issues by building societies.

DEREGULATION

CHAPTER FIVE

The Financial Services Bill

The Financial Services Bill was intended to provide the legislative support for the new markets. Introduced in December 1985, it has subsequently been heavily amended. It is, however, worthwhile examining the original document to see how the legislators saw the investment industry and self-regulatory framework that they intended to impose upon it. For the practitioners there was almost universal dismay. Their reactions towards the Bill are summarised in the following chapter, but the reactions not only of the practitioners but also of the financial editorials were almost uniformly negative.

Outlines of the Bill

Introduction

The Financial Services Bill defines investments and investment business, requires persons carrying on investment business to be authorised and provides for their regulation. It also makes changes to the law on collective investment, listing of securities, offers of unlisted securities and insider dealing.

The goverment's objectives

Efficiency

The financial services industry of Britain should be able to provide services to industry and commerce, private investors and government in the most efficient and economic way.

Competitiveness

The industry must be competitive both domestically and internationally. Regulation must stimulate competition and encourage innovation; it must be responsive to international developments, and not a cover for protectionism.

Confidence

The system of regulation must inspire confidence in issuers and investors by ensuring that the financial services sector is, and is seen to be, a 'clean' place to do business.

Flexibility

The regulatory framework must be clear enough to guide, but not cramp structural and other changes in the industry. It must have the resilience not to be over-run by events.

Transfer of functions

The Bill gives the Secretary of State for Trade and Industry powers to authorise and regulate investment business. Much of the Bill is concerned with these powers, which provide the framework for the new regulatory structure. However, the Bill enables the Secretary of State to transfer the majority of his powers to one or more designated agencies.

Powers can be transferred only if the Secretary of State is satisfied that the proposed designated agency is able and willing to exercise them and that the agency's rules will provide an adequate level of investor protection, will comply with principles set out below, and will not be anti-competitive to any extent greater than is necessary for investor protection.

The chairman and members of an agency must be appointed jointly by the Secretary of State and the Governor of the Bank of England. The transfer of functions will be subject to parliamentary approval.

Although the Bill provides for the possibility of more than one designated agency, it is expected that after Royal Assent, a single body will seek the transfer to it of powers under the Bill.

The Bill leaves many of the detailed requirements to rules and regulations. Most of these will be made by the designated agency.

Requirement to be authorised

Anyone who carries on investment business in Britain must be authorised or, in certain special circumstances, exempted. Doing business without authorisation is made a criminal offence which could lead to fines and imprisonment. The offending business will also be unable to enforce its contracts.

The Secretary of State will be able to apply to the courts for an injunction to prevent a person from carrying on business without authorisation, or for an order (a 'restitution' order) requiring it to repay any profits from carrying on such business and to meet any losses incurred by investors because of the way in which the business was conducted.

There are five main methods of becoming authorised:

Direct authorisation by the agency

The agency will itself be able to authorise 'fit and proper' persons to carry on investment business.

It will be a criminal offence to make a false or misleading statement with the intention of inducing someone to enter into an investment agreement

Membership of a recognised self-regulatory organisation (SRO)
It is expected that most investment businesses will opt to join one of the practitioner-based SROs which will be recognised by the agency under the Bill. To qualify for recognition, an SRO will have to satisfy the agency that its rules and enforcement of those rules provide investors with protection at least equivalent to that provided by the agency.

If a recognised SRO fails to continue to meet the criteria for recognition, the agency will be able to apply for a court order directing it to amend its rules or to improve the enforcement of them so as to ensure that investor protection is brought up to the required standard. Alternatively, the agency will be able to restrict the types of investment business regulated by the SRO concerned or, if necessary, remove its recognition altogether.

Membership of a professional body
Many professionals carry on some investment business incidental to their profession. They are subject to the discipline of their professional bodies, and the Bill will allow these bodies to apply for recognition.

Insurance companies and friendly societies
There is a special regime set up in the Bill for insurance companies and friendly societies.

Businesses from other member states of the EEC
If a business is authorised in another EEC member state which has an authorisation regime providing equivalent standards of investor protection to that in Britain, and does not have a permanent place of business in Britain, then it is authorised to carry on investment business in Britain on a services basis.

Consequences of authorisation

All authorised businesses (except insurance companies and friendly societies) are theoretically entitled to carry on all forms of investment business. However, in practice, all SROs and professional bodies will be required to have 'scope' rules which limit the investment business which their members carry on.

Exemptions from requirement to be authorised
There are certain restricted classes of person who are exempt from the requirement to be authorised. These include public bodies which perform investment business in the course of other duties; the Society of Lloyd's and its underwriting agents as far as investment business undertaken in connection with their insurance business is concerned; and exempt representatives (e.g., 'tied agents') for whom an authorised business takes responsibility.

Recognised investment exchanges and clearing houses
The agency will be able to recognise an investment exchange, if it is satisfied that the exchange's rules and practices provide proper protection to investors whose deals are put through its facilities. It will not be compulsory for all deals to be put through a recognised exchange, but the agency is likely to impose special rules for the protection of investors if they are not.

A recognised exchange will have to demonstrate that it has adequate arrangements for ensuring the performance of transactions effected on the exchange. These arrangements can be either in-house or made through a recognised clearing house.

Conduct of investment business

It will be a criminal offence to make a false or misleading statement with the intention of inducing someone to enter into an investment agreement or to engage in any act or course of conduct which deliberately creates a false or misleading impression of the price.

All investment businesses authorised by the agency will be subject to conduct of business rules. The Bill lays down the powers which the agency will have to make those rules. Recognised SROs and professional bodies will be required to have rules which provide investors with protection equivalent to that provided by the agency's rules.

The agency will be able to prohibit the employment in connection with investment business (except with its consent) of any individual who has shown himself not to be fit and proper to be employed in connection with such business.

If a directly authorised business breached the rules, the powers open to the agency will be:

- the use of its powers of intervention;
- a public reprimand;
- withdrawal or suspension of authorisation; and
- application to a court for an injunction to prevent the breach of a restitution order requiring the business to repay any profits made through breaching the rules and to make good any losses investors have suffered as a result.

Horatio William Bottomley MP, financier, who survived 67 bankruptcy petitions between 1901 and 1905. He was finally convicted in May 1922 of fraudulent conversion and sentenced to seven years in gaol

The regulatory framework

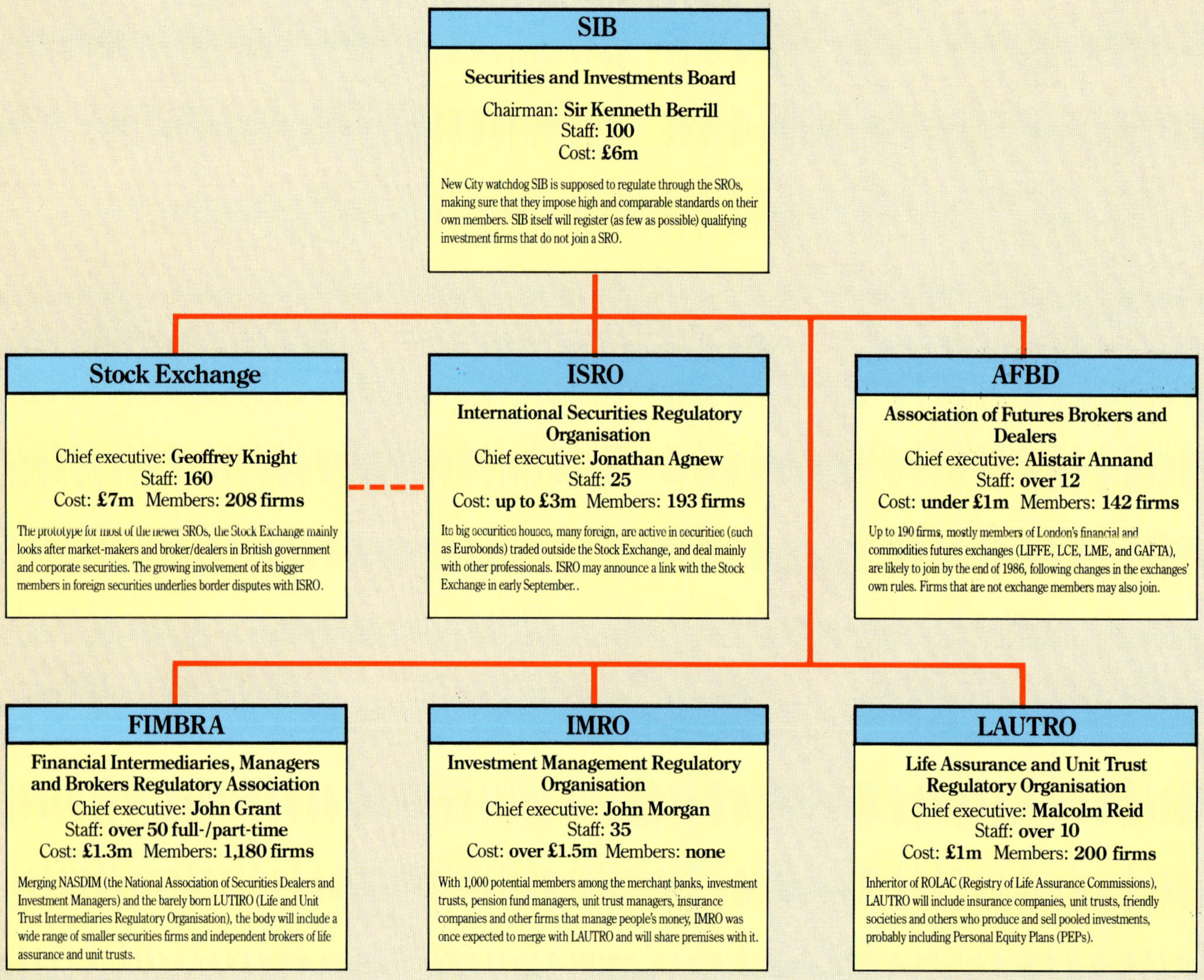

Source: *The Economist*

The Secretary of State will be able to investigate any person who is or appears to be carrying on investment business

In addition, investors who have suffered loss as a result of a breach of rules by a directly authorised business or by a member of a recognised SRO or professional body will be able to bring an action for damages.

Powers of intervention

The agency will have powers to intervene in relation to a directly authorised person to protect investors. These powers include imposing restrictions on the kind of business which may be carried on, requiring assets to be kept in Britain and appointing a trustee to control the assets of an authorised business. Recognised SROs will need to have similar intervention powers, wherever possible in relation to their own members.

Financial service tribunal

If the agency:

- proposes to refuse, suspend or withdraw authorisation;
- proposes to issue a public reprimand;
- uses a power of intervention; or
- proposes to ban an individual from being employed in relation to investment business

the business or individual concerned will have the right to have the case referred to an independent tribunal.

The tribunal will be chaired by a lawyer appointed by the Lord Chancellor and will have two other members appointed by the Secretary of State, one of whom will normally have relevant practical experience. The tribunal will investigate and report on the cases and the agency will be required to implement its decisions.

Investigative powers

The Secretary of State will be able to investigate any person who is or appears to be carrying on investment business (whether or not authorised). The agency will be able to investigate any authorised business. Both will have powers to obtain documents, and to question the business and people connected with it.

Prosecutions

Only the Secretary of State and the Director of Public Prosecutions will be able to prosecute criminal offences under the Bill.

Competition

All rules of the agency and of recognised SROs, investment exchanges and clearing houses will be examined by the Director General of Fair Trading, who will advise the Secretary of State whether they are anti-competitive. If they are, the Secretary of State will have to consider whether the anti-competitive effect is greater than is necessary for investor protection.

Reciprocity

The Secretary of State will be empowered to prevent a financial firm from any foreign country from doing business in Britain (in the banking and insurance as well as the investment sectors), if British firms are not given access to that firm's home market, equivalent to that provided by Britain.

Insurance regime

Insurance companies and insurance intermediaries who sell or give advice on endowment and unit linked policies will have to be authorised under the Bill. For insurance companies, however, there is a special route for authorisation to take account of the existing requirements for the Secretary of State for Trade and Industry to supervise insurance companies under the Insurance Companies Act 1982. The Bill does not disturb this system, which is designed to safeguard the solvency of insurance companies.

Insurance companies will be subject to rules made under the Bill on the marketing of investments – for example, rules on advertising, cold calling and cooling-off periods.

Friendly societies who also carry on insurance business, supervised by the Chief Registrar of Friendly Societies, will be subject to a similar regime.

Independent life insurance intermediaries – brokers, insurance consultants – are treated in the Bill in the same way as other investment intermediary businesses required to be authorised and subject to rules made by a designated agency or SRO.

Collective investment schemes

The Bill reforms the regulation of unit trusts and other kinds of collective investment schemes. Authorised schemes will be those capable of being promoted to the public and thus subject to closest control. Both the manager and independent trustee must, separately, be authorised as investment businesses. There will be powers to make regulations.

The Bill brings together scattered provisions on public offers under one statutory roof.

It is not difficult to see why the Bill aroused such resentment among the practitioners

The Bill replaces the Stock Exchange Listing Regulations 1984 which implemented, on an interim basis, the three EEC Directives on listing. The Council of the Stock Exchange continues to be responsible, as the competent authority for listing, for admitting securities and making rules on listing.

Definition of investments

Under the Bill the term 'investments' will cover:

- stocks and shares in British or foreign companies;
- debentures including debenture stock, loan stock, bonds and certificates of deposits;
- government and other public securities including gilt-edged stock, local authority bonds and bonds issued by foreign governments and international organisations;
- warrants entitling the holder to subscribe for shares or bonds;
- depository receipts for shares, bonds or warrants;
- units in collective investment schemes including units in unit trust schemes and shares in an open-ended investment company;
- option on currency and on any other investment;
- futures contracts for commodities;
- contracts for differences whose value is linked to the value of any kind of property or to an index (for instance, a stock market index); and
- insurance policies which are investments – such as endowment and unit linked policies – but not pure term assurance or policies which simply protect against risk (for example, most permanent health and credit protection policies and general insurance such as fire and theft).

Investment business

A person will carry on investment business if, as a business, he:

- buys and sells investments;
- arranges for others to buy and sell investments;
- manages investments belonging to others; or
- advises others on their investments; or operates a collective investment scheme.

Exceptions

A person who buys and sells investments for his own account will not be carrying on investment business unless he does so to make a market in them, or he solicits people who are not themselves authorised to buy or sell.

Trustees who manage investments in their capacity as trustees will be regarded as carrying on investment business only if they are paid to manage the investments.

A person who gives investment advice purely as a consequence of advice on non-investment matters – e.g. a tax consultant advising a client to sell shares for tax reasons – will not be regarded as carrying on investment business.

Advice included in a bona fide newspaper (but not a tipsheet) will be disregarded.

'Share shops' set up to help employees to exchange shares in the company or group which employ them will be excluded.

Occupational pension schemes

The trustees of occupational pension schemes will be required to be authorised unless management of the scheme's assets is conducted through an authorised person.

Requirements of the rules

The Secretary of State may transfer functions to an agency only if the rules which that agency proposes to make will afford investors an adequate level of protection and comply with principles set out in the Bill. These draw substantially on the existing requirements of the law of agency. They require the rules to:

- promote high standards of integrity and fair dealing in the conduct of investment business;
- make proper provision for requiring an authorised person to act with due skill, care and diligence in providing any service which he provides or holds himself out as willing to provide;
- make proper provision for requiring an authorised person to subordinate his own interests to those of his clients and to act fairly as between his clients;
- make proper provision for requiring an authorised person to ensure that, in anything done by him for the persons with whom he deals, due regard is had to their circumstances;
- make proper provision for the disclosure by an authorised person of interests in, and facts material to, transactions which are entered into by him in the course of carrying on investment business or in respect of which he gives advice in the course of carrying on such business;
- make proper provision for the disclosure by an authorised person of the capacity in which and the terms on which he enters into any such transaction;

- make proper provision for the protection of property for which an authorised person is liable to account to another person;
- make the best provision which can reasonably be made for compensation of indemnity against claims incurred by authorised persons in carrying on investment business;
- require the keeping of proper records and make provision for their inspection in appropriate cases.

Intentions and compromise

It is not difficult to see why the Bill aroused such resentment among the practitioners, even if it is accepted that their opinion represented their own self interest rather than a detached view. There is, however, a widespread view that the new legislation represents a compromise which will achieve very little. The basic intentions were commendable enough. When Professor Jim Gower was appointed by the Department of Trade and Industry in 1981, there was a genuine need for investors to be better informed, even if there was no way in which they could protect themselves against their own gullibility. By 1984, when Professor Gower's report was finished, a fast-changing financial world had already rendered it hopelessly out of date. The inevitable result was that more than 600 amendments were tabled to the Bill, and by the time all parties are satisfied the Bill will bear no resemblance to its original form.

The Bill could not possibly satisfy everyone. The powers of the Securities and Investments Board with the SIB demanding to have absolute authority do cause concern among the self-regulatory organisations (SROs) which are understandably sceptical about reporting to a body which has had limited experience of their business. As the situation now stands, the Securities and Investments Board will be able to make the SROs change their rule books in certain circumstances, but it can only be hoped that the SIB will interfere as little as possible. It does seem an extraordinary decision that the Department of Trade and Industry, rather than the SIB, will be responsible for policing insider trading on the Stock Exchange. This would have provided the ideal opportunity for the SIB to show whether it had any real teeth, a question which, as yet, remains unanswered. Everyone is aware that insider trading is rife, and that the record of successful prosecutions of insider traders is lamentable. Considerable publicity has been given recently to the fact that the number of insider trading investigations has increased substantially, but there has been no proportionate rise in cases being brought to court. The rule-of-thumb principle seems to be that insider trading is almost impossible to detect and that any case of fraud involving less than £100,000 is simply not worth the costs of a trial. In the United States, the SEC has become much more successful at detecting insider trading and, even taking into account the staffing problems of the SIB, it would seem better equipped than the Department of Trade and Industry.

The status of ISRO

One of the major failures of the Financial Services Bill was to recognise the importance and status of the International Securities Regulatory Organisation (ISRO). In essence, ISRO comprises all things 'Euro', which amounts to one of the most thriving businesses in today's financial world. As has been pointed out repeatedly in this book, the new London markets will become almost a mirror image of the Eurobond market. The totally justifiable complaint of ISRO was that the Financial Services Bill clearly shows that the legislators did not understand how Eurosecurities were distributed or to whom they were distributed. Before the appointment of Morgan Stanley International's Archibald Cox to the SIB, ISRO could expect no assistance from that source. ISRO's other dispute lay with the Stock Exchange on the question of 'Euro' and international equities. The Chairman of the Stock Exchange, Sir Nicholas Goodison, clearly had to demonstrate that he was acting in his members' best interests by offering some resistance to the threat that a major part of the Stock Exchange's 'bread and butter' business would move swiftly to the new, mainly foreign-owned, financial institutions. This is likely to prove of little avail in the case of the UK alpha shares, while the British houses have yet to prove that they can consistently win non-British international equity mandates. ISRO successfully argued that its core business was inter-professional and had nothing to do with individual UK investors, who appeared to be the main pre-occupation of Professor Gower's report and of the SIB.

Under the present conduct-of-business rules, life insurance salespersons are still allowed to make random cold calls. Because the sales pitch is usually so persuasive, the result will often be a visit by the salesperson to the home of the prospective customer. And yet, in an interview Sir Kenneth Berrill, the Chairman of the

SIB, openly admitted that some of the worst sales abuses took place when the salesperson had been invited into a customer's home. The life assurance sales industry is highly efficient and, in most instances, thoroughly reputable, but non-professional customers do not always understand what they are buying.

The realities of the market-place

On some points, at least, the SIB has made some sensible concessions to the realities of the market-place. In the Midland Bank Lecture, given at Brunel University, Sir Kenneth Berrill described how he saw his role:

> It has been suggested that more explicit provision should be made for the appointment of independent or consumer members; and I, for my part, would be very happy to see a very significant non-practitioner element on the Board.
>
> But on one point I want to be absolutely clear. No-one on my Board – whether a practitioner or a lay member – is appointed to represent a particular interest or sector of the market. Their background is important for the knowledge and experience that they can bring to bear on the problems that confront us as a Board. In this sense all my Board members must be independent; and they must not, under any circumstances, see themselves as appointed to argue the corner for the interests of their own firm or market.
>
> The Board will receive the statutory powers only if the Secretary of State considers that it meets certain criteria set out in the legislation; these include satisfactory arrangements for monitoring and enforcement of its rules and for the investigation of complaints.
>
> The rules themselves must give effect to certain basic principles set out in the legislation – for example, they must 'promote high standards of integrity and fair dealing' and require an investment business, when acting as an agent, 'to subordinate its own interests to those of its clients'.

Consumers may be ideal for identifying price irregularities in a super-market, but they are not always equipped to give advice on a supposedly hedged commodity fund or a partially-paid Eurobond.

There is little doubt within the government that the Financial Services Bill in its original form was considered to be an excellent piece of progressive and thoroughly prepared legislation. Michael Howard, the junior Minister at the Department of Trade and Industry, said that the Chairman of the Securities Exchange Commission in the United States, John Shad, has told him that if the United States were drafting a new regulatory system, they would use the supervised self-regulation proposals contained in the Financial Services Bill. Michael Howard, by training a barrister rather than a City practitioner, may have been surprised by the controversy the Bill caused and the hundreds of amendments that were subsequently made. However, he staunchly defended the Bill when he was quoted as saying

> The universal reaction in the City is that the Bill's statutory provisions and range of powers are tougher than they expected. In some ways, self-regulation is an unfortunate phrase and some people who work in the City may have been misled into thinking that it doesn't add up to very much. Our statutory framework will be a very effective method of regulation. We're not trying to prevent the City from doing well. We're trying to make sure it does well fairly.

The regulators

Securities and Investments Board	Department of Trade and Industry	Bank of England	Takeover Panel
Securities markets excluding gilts Futures markets Unit trusts Insurance broking Fund management Investment advice Corporate finance advice	Insurance companies Company law compliance including prospectuses, insider dealing, commercial fraud Some responsibility for unit trusts, Lloyds	Gilts markets Wholesale money and foreign exchange markets Bullion Banking	Takeovers and mergers

CHAPTER SIX

The implications of reform

There is no finite solution to the deregulation of the City of London's capital markets. That the London Stock Exchange was falling behind the other capital markets did not require more than superficial analysis. The period between 1979, when foreign exchange controls were removed, and 1983, when the Parkinson–Goodison pact was agreed, provided the ideal opportunity for change. The Stock Exchange could argue that, with legal proceedings under way, their hands were tied. A more critical view is that the Stock Exchange had attempted to form its own 'laager'. The London Stock Exchange has consistently argued that it will retain a reasonable proportion of market share. This may not prove to be the case. The laager – the retreat to within – proved to be no obstacle to the outsiders, whose simple, offensive policy of paying close to the asking price and approximate expenditure of £1.5 billion, has altered the face of the Stock Exchange completely and irrevocably.

There is already an element of doubt about the pattern of ownership in the new London Stock Exchange. The opening of the doors to foreign members was long overdue, but was the invitation intended to lead to a monopoly? The London Stock Exchange, until recently a cartel by any other name, is now facing the realities of overseas competition. The Bank of England's reservations that foreign institutions will dominate the London market will probably be proved correct. The US, European and Japanese houses do not need the laager-concept defence. It will probably take less than two years to identify the weaker members of the new and enlarged market. It is of no importance that some fall by the wayside and others take their place. That will be inevitable within a harsher market environment. From 27 October 1986 the UK houses will find themselves competing head-on with their US, European and other counterparts. There are doubts as to whether all will stand the test, although it can equally be argued that some of the overseas entrants will rue the day that they decided to enter the London markets.

The impending deregulation of London's capital markets brought with it a new sense of urgency to acquire a position within the new framework. The two main jobbers, Wedd Durlacher and Ackroyd & Smithers, were bought almost immediately by Barclays Merchant Bank and S. G. Warburg & Co. The household names in stockbroking fell like nine-pins to domestic and foreign persuasion. With the exception of Cazenove & Co, few houses felt that they were sufficiently strong to remain independent. When the buying rush began, partners of the major London stockbroking firms found it difficult to refuse the financial enticements of a change of ownership when weighed against the risks of pursuing a separate course against financially much stronger newcomers. Both in the

David Scholey, Chairman of S. G. Warburg, who moved quickly to buy jobbers Ackroyd & Smithers

The new structure of the market

In the euphoria which followed the acquisition of almost the entire crème de la crème of London's Stock Exchange member firms came the awareness that profits would be much harder to earn and that the new structure of the market was heavily weighted in favour of the investor, particularly the institutional buyers. Within the City itself, the feeling was one of mixed emotions. While many privately regretted the passing of the old world, the new era appeared exciting. The demand for personnel exceeded the supply. City salaries became a legend, as well as a target for popular condemnation. The practitioners themselves, while leasing new space and installing New York-style trading floors, were carefully watching the passage of the Financial Services Bill through the House of Lords. There were numerous objections to clauses contained in the Bill, and many amendments were made. The Bill, in its original form, had attracted widespread criticism. No-one doubted that it had been drawn up with great thoroughness. It replaced the old Prevention of Fraud (Investments) Act of 1958, a document of 20 pages. The new Financial Services Bill, even without the detailed rules to be finalised later, was over 170 pages.

United Kingdom and overseas, the major financial institutions had to decide quickly whether or not to enter the new London financial markets. For some, including several UK houses, the prices seemed to be too high, and the risks outweighed the potential rewards.

The end-users of the market, such as corporate borrowers and institutional buyers, will have looked at the deregulation measures with a disguised approval. The new legislation was clearly biased in their favour. They would soon be inundated with calls from new market makers offering net prices which would be equivalent to zero commissions in an effort to establish market-share. Corporate funding requirements would move swiftly to the Euro-style where even medium-sized companies would be able to invite bids for a bought deal. Where there were previously three or four financing alternatives, the new capital markets now offered 20. Few would dispute that the overall direction was to a borrower's or buyer's market.

Regulation and the Financial Services Bill

The principal criticism of the original Financial Services Bill was that it had been prepared by non-practitioners and that many of the proposed rules did not take into account the realities of the market-place. In a survey, commissioned by leading accountants Deloitte Haskins & Sells, and conducted by City Research Associates Ltd, 59 banks, securities firms and other institutions were asked for their impressions of the Financial Services Bill as it was passing through its final stages in Parliament. In an accompanying letter dated 24 July 1986, Keith Woodley, the Deloitte partner in charge of the project, made no attempt to hide the problems:

> It is clear from this survey that the organisations interviewed are similarly committed, and are taking measures to secure compliance with the new requirements, recognising that changes are inevitable. It is, however, equally true that there are some aspects of the legislation which have failed to engage their enthusiasm. This is not

Cazenove & Co felt strong enough to remain independent

The principal criticism of the original Financial Services Bill was that it had been prepared by non-practitioners

too surprising in view of the complexity of the proposals in the original Bill and some aspects of the business conduct rules being developed by the SIB which may involve significant changes from current practices. It is clearly going to take time for people to become accustomed to the new framework.

The Survey itself, 'Regulation and the Financial Services Bill'*, was even more revealing as to the general concern of the City. The opinion of the Financial Services Bill itself and its immediate effect was expressed as follows:

The legislation

– Comparatively few believe the legislation as it stands now will do a particularly good job of 'providing a workable framework of regulation for investment business' (19 per cent) or of 'ensuring a level playing field for market participants' (17 per cent).

– Half rate the legislation favourably on investor protection for *private* investors, while less than a third feel the same way about the provisions for *professional* investors.

– A clear majority think there will be no particular benefits to their own organisation (66 per cent) or to their clients (60 per cent) from the new legislation.

– At the same time, 67 per cent do *not* expect the legislation to 'unduly restrict the scope of your legitimate business activities', although around half believe it *will* have a detrimental effect on their relationships with clients (usually in the form of higher costs and additional administration/compliance procedures).

Impact of the legislation

– The vast majority expect the legislation to have a 'significant impact' on the way they conduct their business. Heading the list of individual business areas are market making and broking/dealing (both picked by 47 per cent as likely to be affected), followed by investment management (40 per cent), arranging securities issues (36 per cent) and then corporate finance advice (29 per cent).

– Three-quarters are confident to a greater or lesser degree that they have identified the 'major effects' the Bill's requirements and the SIB's rules will have on their business, although only 29 per cent are 'very confident'.

– Most expect to be able to comply with these requirements 'easily' (29 per cent) or at least 'without any real difficulties' (41 per cent); 14 per cent anticipate problems and the remaining 16 per cent reserve judgement.

Also contained in this excellent study were revealing insights into

Assessing the new legislation

	Excellent %	*Good* %	*Fair* %	*Poor* %	*Not sure* %
Providing a workable framework of regulation for investment business	—	19	41	31	9
Ensuring a level playing field for market participants	—	17	31	36	16
Investor protection for private investors	21	29	29	5	16
Investor protection for business and professional investors	5	24	34	17	19

Source: 'Regulation and the Financial Services Bill', Deloitte Haskins & Sells.

Impact of the new legislation

	Total %	*Commercial banks* %	*Investment/ merchant banks* %	*Securities firms* %
Broking and dealing in securities and other instruments	47	38	67	47
Market making in equities, gilts, Eurobonds	47	46	42	52
Investment management	40	19	92	37
Arranging securities issues	36	23	42	37
Corporate finance advice	29	23	42	32
None of these	14	8	8	21
No impact	7	4	—	16

Source: 'Regulation and the Financial Services Bill', Deloitte Haskins & Sells.

*Available from Deloitte Haskins & Sells.

the approach being adopted by the various institutions towards the new legal framework. For instance, 49 per cent of the firms interviewed had established a compliance department and a further 22 per cent planned to do so. On the question of Chinese walls, 70 per cent believed that it would be possible to demonstrate to the satisfaction of clients or to outsiders that they worked efficiently. Of the various self-regulatory organisations, by far the most popular was the International Securities Regulatory Organisation (ISRO) with 76 per cent of the institutions in the survey having agreed to join. Only 39 per cent were confident that the Securities and Investments Board would be able to maintain its political independence over the long term.

The implications of the Financial Services Bill will clearly affect most areas of business in the new London markets. Perhaps the most surprising statistic in Deloitte's survey was the opinion that 16 per cent of the securities houses interviewed thought that the legislation would have no impact on their activities.

In the conclusion of the survey some of the comments made by the institutions interviewed were clearly critical:

> Badly drafted, ill-conceived. Conflict and confusion. Things have not been sorted out.
>
> The Bill is far too complicated and ill-conceived.
>
> The broadest regulation I have ever seen, the name of the game is chaos and uncertainty.
>
> A poor-quality piece of legislation. It's a mess.
>
> The legislation is muddled. It has been done in too much of a rush and shows a lack of understanding of overseas business.
>
> I have never in my life seen legislation like it. Poorly thought out and a ridiculous hash.
>
> The fundamental flaw is that it is designed for the Aunt Agathas and not the professional investors.
>
> A lot of time and money is being wasted on creating an environment for a number of markets in London, which have operated successfully and without taint, in order to create a perception that the unbridled stockbrokers cannot take advantage of the disestablishment of their privileged cartel.

From these remarks it seems clear that even when the Financial Services Bill has become law, the London investment community will seek many amendments based on the applicability of the legislation to the realities of the market-place.

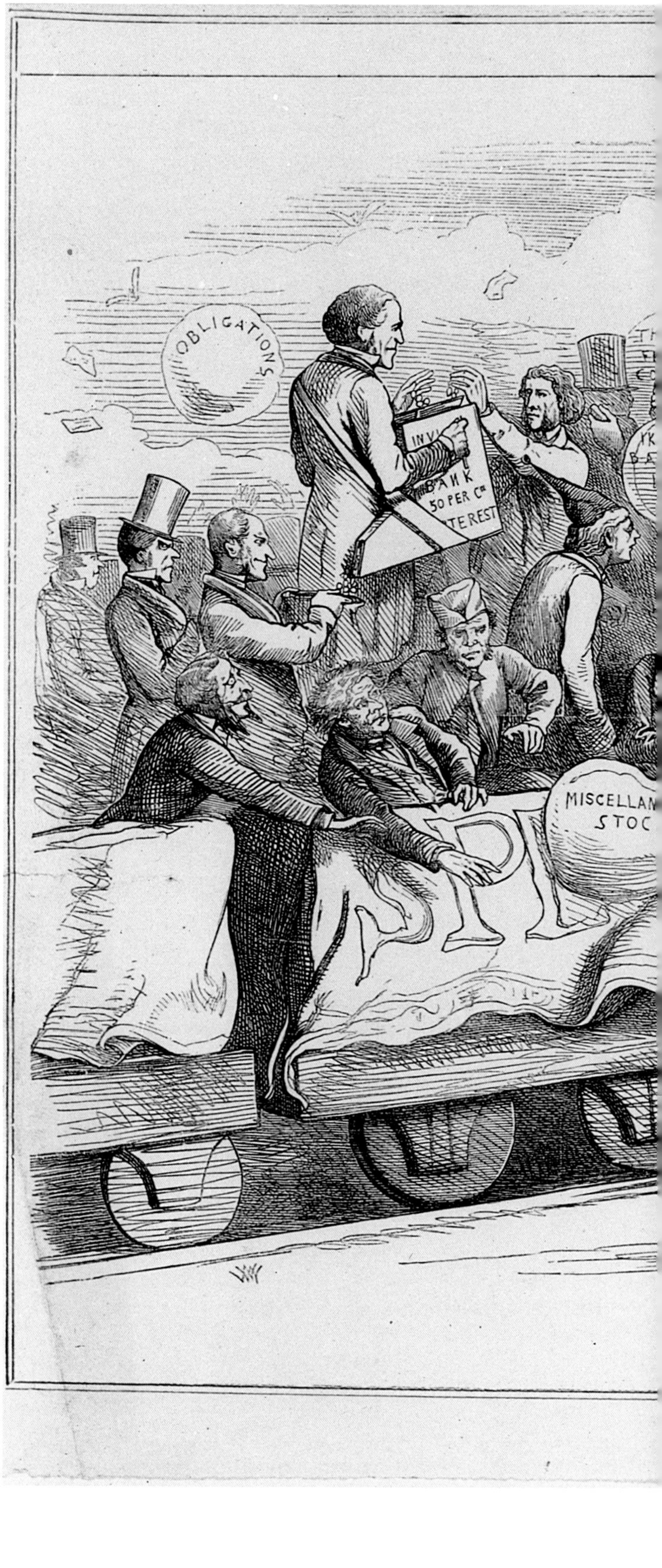

There is nothing new about public fascination and concern about speculation

CHAPTER SEVEN

The gilt-edged market after deregulation

There are few more emotive subjects than the prospects for the UK gilt-edged market or who will emerge the winners and losers in a deregulated environment. In the many interviews conducted in the preparation of this book, normally sedate bankers would become openly irate at the claims of other houses towards minimum anticipated market share. Those houses who had decided not to join the 27 initial primary market dealers (the original number was 31, of which four dropped out) were equally forceful in identifying the dangers of participation.

The charged feelings seem slightly out of character with one of the oldest, and until recently probably the most genteel, capital markets in the world. The traditional image of the gilt-edged market is that of a top-hatted government broker, the senior partner of the former Mullens & Co, walking towards the Stock Exchange. The government has long been the Stock Exchange's largest customer.

Although the size of the UK government bond market is dwarfed by the United States and Japan, it is still the world's third largest domestic market with an enviable reputation for efficiency and dealing size. Transactions of £50 million are commonplace and even orders of £100 million can be completed almost instantly without affecting the price. It is therefore surprising that until 1986 the trading or jobbing part of this multibillion pound market was essentially dominated by two élite jobbers, Wedd Durlacher and Akroyd & Smithers. Between them they held an estimated 80 per cent of the gilt-edged dealing market.

The only certainty of the gilt-edged market after deregulation is that it will never be the same. Wedd Durlacher is now part of Barclays de Zoete Wedd. Akroyd & Smithers and Mullens & Co have been integrated into Mercury International. In their place will be 27 primary market gilt-edged dealers representing some of the largest and most strongly capitalised international and domestic financial institutions. Fourteen are British; the remaining 13 are international, and of these no less than 10 are US-controlled.

Competition

When the Bank of England first asked for primary gilt-edged dealer applications, 31 names were approved. Since this time four have withdrawn: two US names, Bank of America and Drexel Burnham Lambert, and two UK names, Schroder Wagg and Union Discount. Since the first 31 intended participants were named there has been a constant argument concerning the right number, taking into account the present size of the market, the projected increase in dealing volume and the estimated profit to be generated by that volume.

The almost unanimous decision is that there are far too many and that a large number will withdraw after heavy losses and redeploy their capital in more profitable areas. Gordon Pepper, the senior partner of Greenwell & Co (now merged with Midland Bank) and one of the most respected authorities on the gilt market, made no attempt to play down the problems which could occur after deregulation: 'There will be a blood-bath, one-third will have to drop out'. John Barkshire, Chairman of Mercantile House, foresaw 'an enormous free-for-all with no more than 12 to 14 serious players'. David Pritchard, a Managing Director of Citicorp, thought that 'there would be few immediate drop-outs, but eventually the number could be as high as one-third'. Rupert Hambro's comments in a speech made in early 1986 entitled 'Can all 29 primary dealers be right?' were particularly interesting because Hambros Bank, of which he was then Chairman, had taken an early decision not to become a primary dealer.

> Twenty-nine institutions have decided to be primary dealers, and there are several measures which indicate at the outset that this is far too many. The most important measure, I believe, is the comparison with the US Treasury market, on which the proposed gilt market is largely modelled. In the US market there are 38 primary dealers, all of whom have access to the Fed and each of whom is required to conduct 1 per cent of Treasury auction business and regular dealing. But it is a much larger market. Turnover of stock is about US$80 billion per day on a total market of US government debt of US$1.5 trillion. In the United Kingdom market turnover has averaged between £1 billion and £1.5 billion per day over the last two years. In other words, turnover in the proposed gilt market would pro rata represent that of only one of the states of the Union.
>
> A second measure is the likely market for primary dealers' services outside the Stock Exchange. Unlike in the United States market, a significant share of the United Kingdom gilt market still goes to the relatively small holder of stock, who will probably want to continue to use his existing stockbroker. There are only about 150 large institutions actively buying and selling stock. The 29 primary dealers will be competing directly for their business.
>
> A third measure is the general consensus of financial opinion, which is distinctly unhappy about the prospect of the Big Bang and I

quote as my source the poll undertaken by Arthur Andersen which interviewed 600 senior bankers about the forthcoming changes. Seventy-seven per cent of those questioned do not believe that securities trading after the Big Bang will produce returns high enough to justify the large capital put into it by banks. Indeed, I wonder how many of those polled included primary dealers – or what a poll among primary dealers would indicate – because that in fact is the fourth indication that not all 29 primary dealers can be right. The sum of the dealers' own estimates of what their market share will be adds to significantly over 100 per cent. Indeed, nearer 200 per cent. Clearly, not all can be right.

Not everyone, however, was surprised by the number of houses wishing to become primary dealers. Richard Briance, an Executive Director of Credit Suisse First Boston, 'was surprised that the number was not larger'. Stanislas Yassukovich, the Chairman of Merrill Lynch in Europe, said that 'while there may be some drop-outs there would be new names as well'.

To say that one-third or one-half of the primary gilt dealers will withdraw does not take into account a number of factors. First, there is pride. Withdrawal would mean a loss of face and customer confidence. Secondly, there is capital already committed. Apart from the permanent capital required to be a primary gilt dealer, the remaining 27 financial institutions have spent huge sums on building trading and sales teams and installing new technology. Thirdly, the decision to become a primary gilt dealer is often just one part of a much broader strategy. The necessity of immediate profits is therefore not always a consideration.

While it is difficult to imagine that there will be any easy profits for the primary gilt dealers in the early stages, a closer examination of the list of participants does not suggest an immediate contraction of the numbers. All of the US houses have both the capital and the resolve to survive. Those commentators who suggest that Prudential-Bache might be a weak entrant do not take into account the financial backing of the US parent company or the successful acquisition of Clive Discount Company. Among the British houses there are no doubts about the four main clearers or the discount houses. A likely development among the discount houses is that there will be further mergers with larger financial groups. Prudential-Bache acquired Clive Discount, and Alexanders Discount is part of the powerful Mercantile House group. Suggestions that Baring Brothers and Orion Royal Bank may be weak entrants also do not stand up to closer analysis. A strong presence in the gilt market is an obvious extension of Barings' leading position in the Eurosterling and Bulldog markets. Orion Royal Bank may not have the same natural client base in the United Kingdom as some other participants, but Orion has a well-established trading background where determination was underlined in 1974 when the bank continued to trade in the Eurobond secondary market despite sustaining substantial losses.

The starting figure of 27 gilt-edged primary dealers may therefore contract more through mergers than withdrawals. Stanislas Yassukovich's suggestion of new arrivals is also likely to be proved correct. Where, for instance, is Morgan Stanley International, whose claim to be London's largest distributor of international securities would probably be disputed only by Merrill Lynch? The four main Japanese securities houses are almost certain to be entrants at a later stage as they have the wealthiest captive client base in the world. Swiss Bank Corporation is the only one of the three major Swiss banks not to be represented, although Crédit Suisse's participation is through CSFB (Gilts). The role of Credit Suisse First Boston in the new gilts market will be intriguing because of its much envied leadership in both the primary and secondary Eurobond markets. While it may not achieve that

Gordon Pepper, Chairman of Greenwell Montagu: 'There should be more mergers between discount houses and primary gilt dealers'

distinction in the gilt sector, CSFB (Gilts) will be a force to be reckoned with. 'It would not be part of any CSFB future strategy to enter a market unless we intended to be a major factor within that market', said Hans-Joerg Rudloff, CSFB's Deputy Chairman. Another likely candidate for future entry would be Deutsche Bank, whose London-based Deutsche Bank Capital Markets group is extending its presence in a wide range of international capital markets.

The principal change in the new gilts market will not only be in the expanded structure but in the style of dealing. Mercantile House's John Barkshire succinctly summarised the new environment when he said 'positioning will be the key to staying alive'. Miles Rivett-Carnac, a director of Baring Brothers, said 'successful jobbing in gilts will be dependent on successful positioning'. Other bankers are fully in agreement. While the two jobbers concerned declined to comment, one prominent gilt specialist suggested than even Wedd Durlacher and Akroyd & Smithers probably did little more than break even on their gilt dealing books but that they did make substantial profits out of taking a correct view of the market and positioning themselves accordingly. If this was indeed the case it is reasonable to conclude that when two highly professional companies with a near monopoly of a market can make only a marginal profit from the spread between bid and offer prices, it is going to be nearly impossible when the number rises from two to 27. In addition, dealing spreads will almost certainly contract rather than widen.

Expansion of the market

The hope expressed by a large number of gilt participants is that turnover will increase substantially with the arrival of the new primary gilt dealers. If gilts are stable or rising and sterling is not under any undue pressure the trading volume could increase sharply. Some estimates suggest that turnover could rise by two or three times.

In fact, the figure may be even higher as the market moves

Merrill Lynch's Stanislas Yassukovich: 'While there may be some drop-outs, there would be new names as well'

The Hon. George Eden, senior partner of Mullens & Co, government brokers – the traditional image of the gilt-edged market

There is no proven positive correlation between the level of dealing activity and investment performance

quickly towards a US-style of trading where huge positions are taken but the average length of retention is significantly reduced. Higher volume will not necessarily be generated by the customers but rather by the positions taken by the market makers themselves. Rupert Hambro, in the speech already referred to, estimated that there were around 150 institutions actively dealing in the gilt market. Most specialists agree with this figure, although the number of really large gilt players is estimated at between only 50 and 75. Those institutions may well be tempted to increase their gilt trading if dealing costs fall substantially.

In the gilt-edged sector there is little doubt that the arrival of net dealing prices, and the fact that a much larger number of market makers will be soliciting their business will result in lower dealing costs, perhaps by as much as two-thirds for the largest domestic gilt buyers. With this type of cost savings turnover will rise in the initial stages after deregulation. However, as Ron Artus, the investment head of Prudential Insurance who manage more than £20 billion, said in an interview, 'there is no proven positive correlation between the level of dealing activity and investment performance'.

Interviews with a large number of buying institutions and investment divisions of merchant banks made it clear that the institutions would like to maintain existing contacts although they were currently paying indirectly for services and information which they did not always want, and that they would be in a much better position to separate the exact requirements after deregulation. In the gilt-edged market all dealing prices will be on a net basis to the large institutional buyers.

The only reason for not simply taking the most competitive price will lie in the perceived value added by research and other services. Technical analysis has been developed into a state of the art in gilts as well as in US Treasuries, but many UK institutional buyers have developed computer capability which is equal to that of the brokers. A key consideration will be to possess an economist or a team of economists with a proven record of correctly forecasting the market's next change of direction. UK economists have not sought the public recognition of their US counterparts such as Henry Kaufman of Salomon Brothers or Albert Wojnilower of First Boston Corporation. This may change after deregulation. Some US houses who will become primary gilt dealers have already hired prominent UK economists precisely because these economists have direct access to virtually all the major gilt institutions. Their role will be crucial in the new gilts strategy of the foreign entrants and this has been fully reflected in the salaries paid to some of the top names in UK economic forecasting. Not only will the UK economists be encouraged to participate in marketing to institutions, but they will also be the main advisers on the in-house gilt positions. If there will be insufficient profitability in the single spread between bid and offer prices, the success of a firm's gilt operation will have to depend on timely positioning. The associated risks of positioning will increase

Michael von Brentano of Deutsche Bank Capital Markets, a possible candidate for eventual entry into the gilt-edged market

The expertise and sophistication of London's gilt-edged market far exceeds any of its Continental European counterparties

proportionately, but if increased trading volume also means increased price volatility in gilts, there will be a substantial volume of business given to houses whose economists can establish a successful record of short-term forecasting.

The quality of research in the gilt-edged market, already excellent, will continue to rise after deregulation, with Salomon Brothers widely mentioned as being the most serious threat to their UK competitors. For instance, Richard Briance of Credit Suisse First Boston said 'Salomon Brothers are the consumate marketers of fixed-income research in the world'. Other US commercial and investment banking houses are looking at ways in which their new activities in the gilt market will be able to help their global fixed-income, foreign exchange cover and swaps activities. In essence, the gilt presence is only a very small part of a multibillion dollar financial operation. If gilt losses are incurred, the plan is that they will be offset in other markets.

Some UK bankers have said that the US investment banking houses are so averse to losses that they would withdraw from the gilt market if they suffered losses over more than a short period of time. This view may well be wishful thinking on the part of the UK houses. US investment banks have consistently demonstrated their ability to make money in both good times and bad. In addition, the amounts they make are almost unbelievable by UK standards. In 1985 the most profitable US investment banking firm earned profits which probably totalled more than the total profits of all the UK stockbroking firms combined. The same comparison could be made with the largest Japanese stockbrokers, Nomura Securities. It will probably come as no surprise that a significant proportion of these profits were generated by principal trading or positioning for the firm's own account. Positioning is second nature to most foreign houses. Perhaps of even more importance is the amount of risk which they are prepared to assume. If a UK primary market dealer lost £50 million as a result of misreading the market, few would consider the loss insignificant. In the United States, however, losses of US$25 million to US$50 million, although certainly not commonplace, do occur and are accepted as an inherent part of the risk-taking process. Risk taking usually produces profits which allow the US houses to generate earnings which dwarf their foreign competitors. The actual risks are usually minimised by the use of highly sophisticated hedging techniques at which the American houses excel, although these do not always provide the protection that is assumed.

In a recent instance one of the most professional and respected US houses devised a straddle in which a short position in US Treasury bills was hedged by a long position in US domestic corporate instruments. The yield differential between the two was too wide, and as it narrowed to normal historical levels a profit would be secured irrespective of general market movements in the interim period. The exercise did not perform as expected. The yield differential widened rather than contracted and the US house concerned lost an estimated US$50 million for its own account. That it caused no financial concern whatsoever to the US investment banking house and that most of the losses were recouped within two months underlines the different attitudes towards risks which will come with foreign competition in the gilt market.

To justify 27 primary gilt dealers, and quite possibly more at a later date, it is necessary to encourage wider foreign ownership of UK government securities. Recent statistics show that only around 12 per cent of all gilt-edged stock was held by overseas investors. This has to be compared with recent Japanese purchases of US Treasury bonds, which despite the declining US currency may have exceeded US$100 billion or nearly half the amount of the US budget deficit. Japanese institutional interest in gilt-edged stocks has been rising because of the substantial positive yield spread over their own domestic bonds, but it has not been a significant factor as yet.

Continental Europeans have traditionally shown a lack of enthusiasm for gilts because of the overall weakness of sterling since the Second World War and their justifiable suspicion, based on direct experiences of the past, of a future Socialist government. The expertise and sophistication of London's gilt-edged market far exceeds any of its Continental European counterparties. Nevertheless, the ability to execute within seconds trades of £50 million to £100 million, which would need a few days to be completed in some other markets, is no guarantee of investor loyalty. The sharp improvement in the sterling/dollar rate from the low point reached in February 1985 should have initiated a wave of US institutional buying on the basis of equal or better yields plus an attractive currency play, but the actual turnover never reached expected levels. The US institutions should logically be buyers of gilts and the new US gilt dealers will, no doubt, attempt to press home their domestic advantage.

The Japanese should also be increasingly interested in the gilt-edged market, if for no other reason than their sheer volume of surplus funds available for investment. If the projected view of gilts

is perceived to be better than that of US Treasury instruments on a total return basis estimate, the Japanese institutions may be persuaded to switch some of their international investment flows.

The Continental European institutions remain an enigma. They can justify their usual reluctance to buy gilts on the grounds that while sterling has recovered against the dollar, it has been testing new lows against the stronger European currencies, and the benefits of higher current income would therefore have been negated by the accompanying currency loss. The most likely outcome will be that they will adopt an opportunistic buying policy towards gilts rather than accepting the role of a semi-permanent investor. If gilt positions could be satisfactorily hedged in the foreign exchange market against domestic securities, volume could increase sharply. In practice, however, the most likely first generation of matched government transactions will be between gilts and US Treasury instruments or gilts and Japanese government bonds because these are the world's three most efficient capital markets.

Client relationships

Client loyalty will be extremely important in the new gilts market. The attitude of the 50 to 75 major players in gilts may be different from those who rank in the second category. The traditional UK gilt specialist houses such as Greenwell & Co, de Zoete Bevan, Phillips & Drew, Grieveson Grant and Pember & Boyle can argue with some conviction that the UK institutions will remain loyal to the UK houses with whom they have had business relationships for generations. They may be correct and there is certainly a pro-British faction in some of the smaller institutions, including some building societies and local authorities.

However, the understandable wish not to sever established and successful relationships does not take into account three factors. First, institutions are continually being asked to measure their performance against a wider range of indices. Portfolio managers are changed much more frequently than in the past. If performance is to be maximised, it follows that transactions will have to be completed at the keenest prices. Any portfolio managers who do not deal at the best prices will have to convince themselves and their clients that the additional services which they have received compensate for the dealing price disadvantage.

Secondly, the eventual progression of the UK gilt market will be towards global trading. At the present time 24-hour trading is still largely in the imagination, except perhaps in foreign exchange dealing. Within the next few years, however, gilts will not be traded almost solely in London. When the market moves away from London after closing hours, the UK houses will have to ensure that they have a dealing presence in the next phase of the global cycle. If not, a higher proportion of gilt business will fall into foreign hands.

Thirdly, customer loyalty in the gilt market has already been tested by the US giant Merrill Lynch, whose only UK Stock Exchange purchase was a small firm of gilt jobbers, Giles & Cresswell, who previously were probably unknown to the majority

Graeme Gilchrist, Managing Director of Union Discount, who decided discretion was the better part of valour

Patrick Heininger, Director of Barings: aiming for a strong presence in the gilt market

of the international investment community. However, Merrill's strategy was to use Giles & Cresswell as a spring-board into the gilt market by linking the market making capabilities of the gilt jobber with Merrill's huge worldwide distribution system. To strengthen its gilts presence Merrill immediately hired John Hutchinson, a top gilt dealer from Wedd Durlacher. The results of Merrill's entry into the gilt-edged market have been viewed with some concern by their UK competitors, who had thought that the US parenthood would be a detraction in obtaining new business. Previously, Giles & Cresswell had accounted for less than 5 per cent of gilt turnover. Within a few months this share had risen to between 15 and 20 per cent, aided by aggressive marketing in a style which is typically American, i.e. net prices in a larger size and on a lower dealing spread than almost any competitor. Clearly Merrill Lynch will not be able to retain this market share when all the primary gilt dealers are in operation, but it may have demonstrated that price is perhaps a more important dealing consideration than parentage. Accusations that Merrill Lynch 'bought' business at a loss were denied by the company.

The primary dealers

Out of the 27 initial primary dealers, which ones will be power-houses of the future? It is almost an impossible question to answer, especially when bearing in mind that other houses will be joining later, probably at a faster rate than the withdrawals. The most likely shape of the new gilt-edged market will be a tiered system comprising First Division, Second Divison and Third Division. This is not at all dissimilar to the US system where there are 38 primary dealers but, as Jacques Gelardin, Managing Director of Shearson Lehman International in London, pointed out, 'the majority of the business is controlled by no more than eight or 10 houses'. While it would be inappropriate, because the only measures would be largely based on guesswork, to categorise each of the 27 houses, six of the starters emerge as definite candidates for the publicity important First Division bracket. In alphabetical order these would be:

1 Barclays de Zoete Wedd;
2 Goldman Sachs Government Securities (UK);
3 Mercury International (incorporating Akroyd, Rowe & Pitman, Mullens & Co);
4 Merrill Lynch, Giles & Cresswell;
5 Midland Montagu Greenwell;
6 Salomon Brothers (UK).

Some houses will say that this list is incomplete. This may be so and, in any case, the market will change in time. Others will eventually join the premier division, and an initial restrained and cautious approach to the market may prove to be extremely prudent. The selection of the two houses which acquired the pre-eminent gilt jobbers is inevitable if the very substantial prices paid for the jobbers is to be justified. In addition to the unparalleled experience which the gilt jobbers can provide the parents, Barclays

The Bank of England's approved gilt dealers

Aitken, Campbell
Akroyd, Row & Pitman, Mullens (Mercury International)
Alexanders, Laing & Cruickshank
Barclays de Zoete Wedd
Baring, Wilson & Watford
Bankers Trust Gilts
Cater Allen Holdings
Chase, Laurie Milbank and Simon & Coates
Citicorp, Scrimgeour Vickers
County Holdings Group (NatWest)
Credit Suisse First Boston Gilts
Gerrard & National
Goldman Sachs Government Securities (UK)
Greenwell Montagu Gilt-Edged (Midland)
Hill Samuel Wood Mackenzie (Sterling Debt)
Hoare Govett Sterling Bonds (Security Pacific)
James Capel Gilts
Kitcat & Aitken (Orion Royal Bank)
Kleinwort, Grieveson, Charlesworth
Lloyds Bank Group
Merrill Lynch, Giles & Cresswell
Messel (Shearson Lehman)
Morgan Grenfell Government Securities
Morgan Guaranty Gilts
Phillips & Drew Moulsdale (UBS)
Prudential-Bache/Clive Discount
Salomon Brothers UK

Out of the 27 initial primary gilt dealers, which ones will be power-houses of the future?

de Zoete Wedd and Mercury International are generally considered to have applied some of the most thorough long-term planning to the City's securities markets after deregulation. The Greenwell Montagu operation looks particularly strong, and is already a major participant in the market, continuing the financial strength of Midland and the gilt skills of Greenwell. In addition, Midland Bank recently announced that it would create it own discount house through Greenwell Montagu Gilt Edged Ltd. This development would reinforce Gordon Pepper's argument that 'there should be more mergers between discount houses and primary gilt dealers'.

The selection of the three US houses is based on capital backing, aggressive marketing skills, a proven willingness to take enormous risks and a commitment to become a major participant in all the world's deregulated government securities markets. Firms such as Salomon Brothers, Goldman Sachs and Merrill Lynch are simply not prepared to participate in markets in a Second Division category and history suggests that they do not generally prefer a gradual approach.

What about the composition of Division Two? It would take a brave man to draw up a list because of the offence it would cause to those houses which are omitted. It should not be implied that because a UK house or affiliate has not immediately been nominated for the First Division category it has no aspirations towards that status. County Bank has the balance sheet of National Westminster Bank behind it, but it does not have the gilt skills of, say, Greenwell. James Capel and Hoare Govett have excellent

CSFB's Richard Briance: 'Surprised that the number of primary gilt dealers was not larger'

David Potter, Managing Director of Samuel Montagu: Midland has created its own discount house through Greenwell Montagu Gilt Edged

reputations in gilts, and wealthy foreign owners who will assist them in maintaining research and services to the highest possible standards, but they still may find themselves losing market share to Salomon Brothers, Goldman Sachs and Merrill Lynch. Phillips & Drew will be hoping that Union Bank of Switzerland's legendary client base will be persuaded to include gilt-edged securities in every representative portfolio, but it will have to overcome the longstanding suspicion of sterling as a currency. Kleinwort Benson and Morgan Grenfell clearly intend to be very aggressive in all aspects of the financial services industry and they are definitely candidates for Division Two.

Among the US houses, Shearson Lehman would almost be a Division One contender in its own right, and Bankers Trust and Citicorp are simply too powerful to ignore. Credit Suisse First Boston remains something of an enigma, but for the time being it may prefer to retain its premier position in Eurobonds and Euro-equities. The Division Two table, again in alphabetical order, would probably include:

1 Bankers Trust Gilts;
2 Citicorp, Scrimgeour Vickers;
3 Kleinwort Grieveson;
4 Morgan Grenfell Government Securities;
5 Security Pacific, Hoare Govett;
6 Shearson Lehman, Messel;
7 UBS, Phillips & Drew.

The US newcomers will transact a large portion of their own in-house business through the interdealer brokers authorised by the Bank of England. In the United States the domestic bond brokers are an integral part of the US government bond trading system. Despite considerable opposition the bond brokers now play a crucial role in the Eurobond market, where they were initially considered to be no more than parasites taking business away from the recognised market makers. But they have now been recognised as intermediaries without which the primary market – which in 1985 raised the equivalent of around US$133 billion – would be unable to function as efficiently as it does. The degree of anonymity provided by the interdealer brokers will enable houses to move large gilt positions for themselves and their customers. The profitability of the Eurobond prompted Rupert Hambro to say that 'given freedom of choice, it might be relatively more rewarding to be an interbroker dealer than a primary gilt dealer'.

Precise market share is a subject that most of the 27 primary gilt dealers are now reluctant to quantify. Gone are the days when the then 31 likely primary market dealers all estimated 'a minimum market share of 10 per cent as being the initial objective'. A far more realistic view was given by Mercantile House's Chairman John Barkshire, who said that 'a market share of 7.5 per cent would be realistic within a three- or four-year period, although gilts form only a relatively small part of our overall expansion strategy'. Some powerful UK participants have also changed their original objectives of market share in the gilt sector. In the *Financial Times*' edition of Wednesday, 19 December 1984, Michael Hawkes, the Chairman of Kleinwort Benson was quoted on the bank's future role as a primary market dealer as saying 'we will do as little as possible'. In a more recent interview Kleinwort Benson 'expected to maintain their present share of the gilt-edged market of around 10 per cent'. The change may have reflected Kleinwort Benson's emergence as one of the most powerful and innovative UK merchant banks who see no reason to surrender any domestic territory to the foreign newcomers in the gilt-edged market, and have been among the first to carry the fight back to the domestic preserve of the US invaders.

Conclusion

It will take several years before we know who will be the winners in the gilt-edged market. In the early stages it is probable that the London houses will have to cede a large part of their previous dominance to foreign institutions, and in particular to the Americans. By 1990 foreign-owned primary gilt-edged dealers will probably control more than 60 per cent of the gilt-edged market. The ratio of new entrants into the gilt-edged market will also be heavily weighted in favour of foreign firms. The one compensation for the UK houses is that future entrants will have to hire London-based staff and expertise in order to operate efficiently and profitably.

CHAPTER EIGHT

The equity markets after deregulation

There were several prominent City bankers and financial specialists who had the vision to see that the former structure and protectionist attitude of the London Stock Exchange was hopelessly outdated and would eventually have to change. One of the earliest claimants to have planned a strategy for a deregulated market was Mercantile House's John Barkshire, who discussed the purchase of a UK securities house as long ago as 1967. Later Mercantile House became one of the first UK groups to own a major Wall Street firm when it acquired Oppenheimer & Co. Gordon Pepper of Greenwell & Co foresaw deregulation 10 years before it was put into effect. Michael Hawkes of Kleinwort Benson had meetings with a leading stockbroker and a smaller investment bank when UK foreign exchange controls were removed in 1979. Dean Lundell of the US bank Security Pacific, whose initial purchase of a minority stake in Hoare Govett marked the first acquisition of a top-ranking UK broker, traces the origins back to the early 1970s. He gives credit to Frank Cahouet, then head of Security Pacific's operations in London before returning to California, where he finally moved from Security Pacific to take over briefly the troubled Crocker Bank before it in turn was absorbed by Wells Fargo Bank. 'Frank had been involved in investment banking himself and for a long time he harboured the wish to own a merchant bank in the United Kingdom', said Lundell. He added, 'when the regulations were changed our natural choice was Hoare, Govett & Co with whom we had a long business relationship and who had first listed our shares on the London Stock Exchange.'

However, while others talked around the general subject, some decided to try and capitalise directly on their vision of the future. Among the first was Jacob Rothschild, widely regarded as a maverick by some of the City establishment, but admired by all for his dealing skills and for his ability to be able to identify trends long before they were recognised by others. Jacob Rothschild, together with colleagues Rodney Leach and Philip Shelbourne (now Sir Philip), had propelled N. M. Rothschild & Sons into the forefront of the Eurobond market in the later 1960s. Their deals had included some of the most innovative deals of the time, including the tranche series of Trasalpine Finance loans, which may have been one of the first instances where a project finance deal was effectively 'securitised' by transforming the loan into a public bond offering. N. M. Rothschild, S. G. Warburg and Hambros Bank were the three UK merchant banks which played an effective role in the early growth of the Eurobond market. Later, when Jacob Rothschild had

Jacob Rothschild sensed the impending changes long before the Parkinson–Goodison agreement in 1983

The London Stock Exchange prior to Big Bang's changes

left the family bank whose Euro aspirations had by then waned, he chided the UK merchant banking community for 'handing the Eurobond market to the Americans on a plate'.

Jacob Rothschild sensed the impending changes of direction in the London Stock Exchange long before the Parkinson–Goodison agreement in the summer of 1983. Through Rothschild Investment Trust he had bought minority shareholdings not only in Sotheby, but also in the leading jobbing firm of Wedd, Durlacher, Oldham and Mordaunt, now wholly owned by Barclays de Zoete Wedd. It could be argued that Jacob Rothschild sold both stakes too early, but when he subsequently bought a minority interest in the small London stockbroking firm Kitkat & Aitken (since sold to Orion Royal Bank), he again highlighted the Stock Exchange's major problem – inadequacy of capital. The combined capital of the London jobbing firms amounted to less than US$150 million compared to the US$2 billion each for publicly-held Merrill Lynch and American Express Shearson Lehman. The total profitability of all the London Stock Exchange firms amounted to less than the 1985 earnings of Goldman Sachs which was not even publicly owned.

Why then should outside institutions, including a high proportion of foreigners, have stampeded to buy London stockbroking firms which were poorly capitalised, modestly profitable and often not very well managed? The answer is that many felt there was really no other alternative. London was destined to become one of the world's major financial centres and the price of entry was bound to be high. Nevertheless, the speed at which Stock Exchange companies were acquired surprised almost everyone. Stuart Valentine of the Stock Exchange said, 'Yes, the speed of the mergers almost implied an element of panic', and he added, 'the mergers which were arranged in haste could go wrong'. The prices paid for the London houses drew considerable criticism from outside the City, an opinion which was no doubt at least partly motivated by jealousy. Securities houses are not easy to value because they have few fixed assets, and a high proportion of the price paid must inevitably represent goodwill. 'We probably all paid too much for mergers', said Orion Royal Bank's Philip Hubbard and Michael Perry.

However, on closer examination this may not always prove to be the case. The minority interests acquired in 1983–85 have in many cases probably been excellent investments due to the strength of the London markets and record trading volume. Security Pacific's

Most estimates for the United Kingdom expect a decline in equity commissions of between 30 and 40 percent

acquisition in 1982 of 29.9 per cent of Hoare Govett has unquestionably been a superb bargain.

Another factor which non-practitioners have failed to take into account is that the acquisitions were made by highly professional institutions with particular expertise in the pricing of financial services. They probably never expected to find a bargain in the same category as Security Pacific's acquisition of a stake in Hoare Govett, but neither were they prepared to offer absurdly high prices. Far too much has been made of the City's new 'millionaires club'. Several hundred men and women can now be included in this category, but much of the new wealth is in the form of paper rather than direct cash and is usually being paid in instalments to ensure that the individuals concerned remain with the organisation. In addition, a substantial part of the sums received will end up with the Inland Revenue.

The detractors should also remember that holdings in a stockbroking firm were almost totally illiquid and that the industry itself was notoriously cyclical. In the worst years the partners received nothing, and indeed occasionally they had to inject personal funds to provide adequate working capital. In the case of a London stockbroking firm failing to meet its obligations the partners of the firm were liable for their entire personal assets. Critics of the high rewards in the industry often overlook the high risks which accompany the rewards.

The concern of the acquisitors was probably not so much the cost of the actual purchase, which in most cases was insignificant in relation to their overall capital, but the future value of their new investment. Very few of the acquisitors expected a high immediate return on capital. When asked by an interviewer where the profits were to be derived, Stock Exchange Chairman Sir Nicholas Goodison replied, 'Where indeed?', and this reply must have echoed in the minds of the buyers. There will be no easy profits. Any institution which has decided to be a major participant in the new deregulated London market and which is not prepared to take a two- or three-year view will probably be sorely disappointed.

The US experience

The attitude towards the market in London should be the same as it was in New York when fixed rates of commission were abandoned in May 1975. Rates of commission did fall very sharply, although in the initial stages discounts were mainly confined to institutional orders of US$300,000 or more. The subsequent free fall in New York taught three important lessons, which it is hoped will not be lost on London.

First, client loyalty is fickle. The US brokerage houses who had relied upon, and who may even have been promised, guaranteed levels of business by the clients found that the clients did not always keep their word.

Secondly, research is an extremely difficult commodity to value. Almost all the leading research boutiques had either failed or had been absorbed into larger Wall Street houses within a few years. Classic names such as H. C. Wainwright & Co, Baker Weeks & Co and Spencer Trask & Co were among the casualties as it was realised that, while institutions still wanted the research, they were not prepared to pay very much money for it. UK institutions have generally agreed that they will continue to value high-quality research, but the New York example is not very encouraging. The large institutions will probably be very ruthless when they address the question of 'unbundling' of services which they receive from their stockbrokers. No doubt many feel that they are receiving a great deal of information which they do not particularly want. This will be eliminated in the unbundling process, and the institutions may try to renegotiate the total levels of commission paid to the reduced amount of information which they are receiving.

Thirdly, the unbundling of stockbroker services will create a number of discount brokers on similar lines to their no-frills counterparts in the United States. The discount brokers will essentially provide no other service than execution. There are already a number of fringe discount institutions in the United Kingdom which have been relatively successful despite being undercapitalised and having less than blue-chip reputations. Traditional city stockbroking houses have expressed reservations concerning the need for discount brokers, arguing that the retail customer base in the United Kingdom is limited.

This view is unlikely to be correct. In the United States discount brokers have prospered, with firms such as Charles Schwab and Quick & Reilly proving to be very successful. At some point one of the UK clearing banks might well purchase a discount stockbroker, thereby emulating Bank of America's acquisition of Charles Schwab, which has proved to be so profitable. In the United Kingdom smaller retail clients traditionally deal through their local branch banks, which is extremely expensive and helps to pay for the research

Launch of the British Telecom prospectus. From left to right: BT's chairman Sir George Jefferson, David Clementi, Sir Martin Jacomb and Lord Rockley

which the stockbrokers provide to the institutions rather than retail clients. The smaller clients would therefore no doubt welcome an opportunity to reduce their dealing costs by probably 50 per cent in situations where they do not require any additional specialist advice.

Commissions

The investment community, particularly the institutions, stands to gain substantially in the new, deregulated market. They will be able to pick and choose the services which they require from the broking community. They will be offered net prices in 100–150 of the leading UK shares. In many instances the best regarded analysts will also have traders who are market makers in the securities which they cover. Gilt markets will move immediately to a net price basis. The arrival of discount brokers will enable institutions to cut costs in stocks where they have reached a conclusion independent of stockbrokers' analysis. To what extent will commission fall?

After fixed rates were abandoned in New York in 1975, the initial decline was no more than 10 per cent but accelerated very soon afterwards to 40 per cent. Most estimates for the United Kingdom expect a decline in equity commissions of between 30 and 40 per cent. In April 1975 a Wall Street bargain of 1,000 shares at US$100 per share attracted commission of US$0.75 per share. Today the

negotiated commission would probably be in the region of US$0.15–0.20 per share, and certain US institutions would demand even finer terms. This suggests that equity commissions could fall by 50 per cent or even more. In a recent article a major investor, County Bank, was quoted as saying that it currently paid an average commission of 0.25 per cent and expected to be able to negotiate a flat rate between that figure and the 0.125 per cent charged by Ariel, the computer-based service owned by a number of institutions which matches buy and sell orders. In the select group of top-ranking shares, which are traded on a net market basis, it is difficult to see the major institutions passing orders on an agency basis which includes the payment of a commission.

If Stock Exchange commissions do fall by 50 per cent or more, will a number of member firms not fail? Certainly there will be a series of mergers among the smaller companies, and more regional groupings among the country brokers who should pool their resources of local knowledge to offer their clients wider coverage. This has already started to take place and will be emulated by other country firms. The smaller organisations with a good client base will be less affected by falling commission rates. Regional loyalties are strong and there is less reliance on institutional business. 'Some country firms have never seen an institutional order', explains the Stock Exchange's Stuart Valentine.

The smaller firms based in London with a mixed business could prove to be vulnerable. London overheads are rising continually. Star employees are bound to be tempted by the greater job security and rewards offered by a larger firm owned by a bank with virtually unlimited financial resources. Private clients of smaller London firms may appreciate the personal service which they receive, but they will also insist on checking that transactions are executed on the 'best execution basis' which may entail significantly lower commissions. It was therefore surprising to read statements from some Stock Exchange Council members that private client commissions could actually rise. Certainly, private clients will never be able to demand the type of terms asked by the major institutions, but they will always have the alternative of dealing through the discount brokers. Many smaller private clients have accounts with stockbrokers, but only a minority are handled on a fully discretionary basis and many of the clients prefer to make their own investment decisions.

The most encouraging aspect for the smaller stockbroker is that the US and Continental banking institutions who have bought UK stockbrokers have very little interest in the smaller UK investor. Even retail-oriented houses such as Merrill Lynch have made no attempt to widen their share of the UK retail market, preferring clients with a minimum portfolio of £100,000. Conversely, the main opposition to the small stockbrokers may come from organisations such as the major insurance companies and other institutions, who will be mounting a campaign to sell the retail client every financial service from endowment policies to mortgages to car loans and, of course, investments in the form of unit trusts.

Chinese walls

It is unclear at this stage whether the Stock Exchange will function more efficiently for investors, without whom it cannot survive, or for itself as a capital-raising mechanism. Investors know that commissions are coming down, but will the service be as good as before? Some institutions are openly sceptical about the effectiveness of Chinese walls when analysts and net market

Richard West, Deputy Chairman and Chief Executive of Hoare Govett. Security Pacific's early acquisition was said to be 'top quality for a bottom dollar'

Some institutions are openly sceptical about the effectiveness of Chinese walls when analysts and net market makers may be sitting side by side

makers may now be sitting side by side. Who receives the news first if a top-ranked analyst switches a recommendation from a buy to a sell? Every bank will say that it is the clients, but if the net market maker is already long of, say, 500,000 shares at a higher price, he or she will not be overly pleased to be deluged with a wave of sell orders.

Almost every bank has admitted that they will base their net market making in equities around the companies which are covered by their research analyst. In these circumstances it may be extremely difficult to avoid a conflict of interest. Up to this point traders and analysts or economists have rarely worked together, although such a situation is common in New York. The US houses contend that conflicts do not exist, a view generally shared by the clients. Nevertheless, in an industry which will be dominated by the profits generated from positioning as well as commissions, the trader would clearly be interested in any changes in the in-house analytical opinion.

The separation of investment management from the sales, research and trading areas appears to be one of the lesser problems in the post-deregulation period. Satisfactory Chinese walls have been in place for some time, and fund management operations have often been removed to totally different premises. The top stockbroking firms such as Scrimgeour, Phillips & Drew, James Capel, Cazenove, Wood Mackenzie and Hoare Govett have built up very substantial private client businesses without having to seperate them from the research departments or abusing their role as official brokers to many hundreds of companies.

A more serious question arises when major financial consortia are formed which incorporate merchant banking, research, net market making, equity sales and distribution. A non-City practitioner would find it implausible that these various groups do not exchange views on their existing, or targeted, corporate clients or their subjects of analysis. All banks will assure their customers that their Chinese walls not only exist, but that they are totally effective. In a commentary entitled 'Building up a major group', John Quinton, the Deputy Chairman, and Geoffrey Guinness, the Assistant General Manager of Barclays Bank, highlighted the attitudes towards Chinese walls and conflicts of interest.

> Both the Stock Exchange and the Department of Trade in its White Paper on investor protection have proposed codes for the regulation of conflict. We certainly believe strongly in the overriding need for an agreed code of conduct, which will be observed by all parties to the market. BZW itself will have a code of practice for its staff, which will incorporate the best market practice together with any code of conduct agreed at the Securities and Investments Board (SIB) or Stock Exchange level. Furthermore, in accordance with US practice, BZW will establish a compliance department staffed at the highest level which will have the responsibility of ensuring that in all situations where conflict could arise, appropriate procedures for the protection of clients have been observed.
>
> People have said to me that Chinese walls will not work in a large group like ours – that when the banking side, for instance, learns that a company is running into trouble it will mark the card for the market makers and brokers in BZW. We cannot emphasise too strongly that this must not and cannot be allowed to happen and that rigorous internal action (such as a fine or even dismissal) would be taken if it did. In such circumstances the offending group might also be liable to a penalty imposed by its SRO [self-regulatory organisation] or the SIB. We hope the range of their penalties would not be limited to censure or suspension – the two extremes – but would include the power to impose a monetary fine, which might well be a more appropriate penalty, not only for a group under an SRO but also for a firm or individual authorised by the SIB.

In practice the system has worked remarkably well in the United States, where investment banks have been fulfilling for years all the functions which the London consortia will now offer. A salesperson in Los Angeles has no inkling of the activities of the New York-based corporate finance department. Often the first news that the trader will hear is that a particular stock in which his firm makes a market has been placed on the restricted list, which effectively precludes dealing. Some have taken the conflict of interest subject so seriously that they have considered not making markets at all in the shares of companies with which they have a traditional advisory relationship. This is well-intentioned, but is also probably impractical. Traders tend to deal on free market forces, and just because they adjust their prices to supply and demand does not imply the possession of privileged information. Also, a trader may make a market in a typical top-ranking UK company while his bank may be advising the same company on the feasibility of issuing Eurocommercial paper. The Eurocommercial paper issue, if completed, would be unlikely to affect the company's share price, but the bank is still an official adviser. Should the shares be placed on a restricted list? Probably not. The restriction would only apply if

The principal threat to the UK stockbroking community from the overseas newcomers is in the area of Euro-equities

an announcement, such as a takeover, was about to be made which could materially affect the company's share price. Even in this case the liquidity of the underlying share could be seriously impaired if the trader was also by far the most important market maker.

There is no easy solution to this question, but it seems reasonable to suggest that there will be a sufficient number of market makers in the alpha UK stocks to ensure that reasonable liquidity is maintained in the event of one or two market makers being restricted in their dealing.

The next logical question is what would happen in a similar situation to the beta and gamma stocks which will be traded by many fewer houses? The likely answer here is that they will only be a handful of net market makers in the beta and gamma categories and that almost all business will be transacted on an agency basis rather than a net basis. Dealing spreads will almost certainly be wider, but this is no different from second and third stocks in any other equity market.

A number of the most highly regarded UK stockbroking firms have chosen slightly unorthodox approaches to the deregulated London stock markets. Cazenove & Co, widely regarded as having the best placing power in London, chose not to team up with any major financial institution despite probably being at the top of everyone's shopping list. 'Cazenove doesn't need to join forces with anyone', said the senior partner of one London stockbroking firm, 'they have always acted like a merchant bank so why do they need to join one?' Another partner at the same firm said 'Cazenove have never tried to be number one in research but they are brokers to most of the best companies in the UK, they have the best entrée into the investment trusts and they are brokers to the majority of the most successful new issues in any one year. The result is that very few institutions are able to turn down a Cazenove offering.'

With such a dazzling reputation and corporate client connections it is perhaps not surprising that Cazenove has essentially elected to continue its success through conventional commission-based business, corporate finance activities, portfolio management and venture capital projects. By remaining independent it may actually increase its number of corporate customers, some of whom have openly expressed reservations about their former stockbrokers being merged into an organisation with whom they have no previous relationships. Some UK companies will have found that their traditional brokers have been absorbed by a financial institution which previously represented a client, who had unsuccessfully attempted to acquire them. In London's deregulated world it is not only customer client loyalty which will be tested, but also corporate client loyalty.

Euro-equities

The principal threat to the UK stockbroking community from the overseas newcomers is in the area of Euro-equities. Although promoted several years ago by far-sighted bankers such as Stanislas Yassukovich, Chairman of Merrill Lynch (Europe), the concept has really only become a fact within the last two years. Previously global distribution was, in effect, an outward attempt to denote a group of international underwriters who subsequently sold their stock back into the domestic market-place. Placing power had not been established. Research capabilities were non-existent. Genuine international asset allocation was in its embryo stages. The US institutions had only between 2 and 3 per cent of their portfolios invested abroad.

With increasing international investment understanding and the superior performance of many overseas markets, the Euro-equity has come of age. Why is the description 'Euro-equity' preferred to 'international' equity? The answer is that the distribution system is based on the Eurobond market using an international rather than a domestic syndicate. This can be further refined by geographically dividing the responsibility for sales between specific regions. For 'jumbo' equity issues such as the 1985 flotation of British Telecom, or for a major secondary offering such as British Petroleum, the benefits of global distribution were recognised by the international investment community.

The system was then, and is still now, far from perfect. The natural gravitation back into the domestic equity market was clearly visible. The international ability to place into firm hands was therefore always in some doubt. Considerable prestige was attached to appearing on the front page of the offering prospectus. Some underwriters therefore promised far more than they could actually deliver. In the case of British Telecom, the US underwriting syndicate attracted widespread criticism for selling shares almost immediately back into London, thereby substituting genuine distribution over a gradual period into a quick underwriting profit. There is no foolproof method, as yet, of ensuring that an international underwriting syndicate does not take the easy option of selling back into the domestic market. In the Euromarkets this has been assisted, to a major extent, by the anonymity provided by the bond brokers.

In a recent speech the Chairman of Swiss Bank Corporation International, Andrew Large, succinctly defined the major parties in a typical Euro-equity transaction and the different problems which they were required to address.

> First, the intermediaries: domestic equity issuing syndicates have, of course, grown up in all the main industrialised countries. They've tended to be rather clubby. Intruders haven't, on the whole, been welcome. The merchant bankers and brokers looked after their clients and . . . I think its fair to say . . . often looked after their own interests pretty well, too. The system has basically worked, so there was little incentive to change it.
>
> Second, the shareholder institutions. The big institutions will, on the whole, want protection to avoid new shareholders diluting their percentage interests. They haven't exactly welcomed foreigners coming in and offering to pay cash for new shares at or around the market price. And they've wanted the *benefits* of pre-emption rights. Nice discounts. A cynical comment, maybe, but a true one.
>
> And the third party – that's the issuers. The issuers have basically done what they're told. There just didn't seem to be an alternative to the traditional domestic syndicate. Amounts have been easily absorbed in the main markets. And on top of this they've often been influenced by political or other considerations. Such as trying to keep the shares of a country under ownership of its own nationals. One sees that pattern in defence contracting, and other fields in different countries.
>
> The net results of all this was, in the past, that it was easy to dismiss the Euro-equity market. As I said earlier, you can't afford to make mistakes with equity issues. And there has clearly been plenty of ammunition supporting the view that there's really no need to experiment. We've had the old problem of flow-back. You may remember the first early attempts at arranging international equity offerings. They were disastrous. In the early 1970s several companies, like Baxter Travenol of the United States and J. Lyons in the UK, attempted to place equity internationally. The deals were all flops because all the shares flowed back to the domestic markets. Those events gave rise to feelings that all the foreigners had wanted to do was to pocket the commissions.
>
> And it's not hard to find other examples. Sometimes the regulatory environment has been used as a means of preventing change. In the past, for example, the strict adherence to the principles of pre-emption rights effectively prevented new issues for cash taking place; other than as rights issues. Not that I am advocating a 'Wild West' with respect to pre-emption rights. But, taken to the extreme, there is no question that if issues are only tailored to existing shareholders, then this must cut off potential sources of capital: that could be of benefit to the issuers and their shareholders.

Andrew Large, Chairman of SBCI: 'Intruders haven't, on the whole, been welcome'

By the standards of its big Eurobond brother the Euro-equity market is miniscule. From the earliest issues of European Depository Receipts back in the 1960s, the market was producing little more than US$90 million equivalent in 1983 and totalled only US$2 billion equivalent in 1985. Nevertheless, volume will continue to grow, a process that is already certain due to the several privatisation programmes under way. The privatisation programmes in the United Kingdom and France will confirm that even if domestic markets are large enough to absorb the privatisation offerings, they are not necessarily the most efficient or productive means of distribution. Euro-equity offerings may not achieve the grand objective of global distribution but they will result in wider share ownership. For the larger UK companies Euro-equity offerings such as the recent offering by Hanson Trust plc do not signal a move directly away from traditional City financing, but they do demonstrate that corporations now have an attractive and viable alternative.

Hans-Joerg Rudloff, Deputy Chairman of CSFB: 'It would not be part of any future strategy to enter a market unless we intended to be a major factor'

Most of the major participants in the new equity market agree that success will be based on a combination of market making skills, sales distribution and research. There will be profitable niche areas in sectors such as smaller private client business and the Unlisted Securities Market (USM). However, these remain of secondary importance in the period immediately after deregulation, when all attention will be focused on the institutional business. The institutions will be in a position to call the tune because they will be inundated with houses aggressively soliciting their business. They will be offered net market prices on the lowest spreads and in the largest size they will probably have encountered. With the exceptions of élite houses such as Cazenove, with its stable of corporate clients, and James Capel, with its undisputed excellence in research, the medium-sized firms relying on agency commission business may see their market share decline sharply.

The new equity rules are not numerically different to that of the gilt-edged market, where previously there were two major market makers and now there are 27. In the equity sector there were previously around seven leading jobbers, but in the new market place there will be around 60 houses making net prices. The result is that the cake either will have to be cut into many more slices, or will have to expand very substantially. The final outcome may rest on the investor sentiment towards the London equity market and sterling in the period after deregulation. The worst possible situation would be for London to be in the early stages of a major bear market.

THE INSTITUTIONS AND THEIR STRATEGIES

Right: John Barkshire, Chairman of Mercantile House, foresaw the coming of deregulation

Below: David Benson, Director of Kleinwort Benson responsible for international capital markets, and, below right, Lord Rockley, Vice-Chairman of Kleinwort Benson

CHAPTER NINE

The new British financial institutions

In interviews with, or in comments made by chairmen or senior executives of the new British groups, all profess to have a clear, well-defined strategy for the deregulated London markets. The truth is probably that the strategies which reassure the public as well as the shareholders, are more of an exercise in hope. It is impossible to formulate a precise strategy for a deregulated market which is a completely unknown environment.

The new British consortia now preparing to defend their domestic territory are fully aware of the competition they face. Many feel that the deregulation process happened too quickly. Fred Carr of Capel-Cure Myers in an article wrote: 'Did the government really intend to create a market likely to be dominated and perhaps controlled by foreign institutions? It isn't what they said they were trying to do, but it may well happen.' Although the marriage process was in some instances haphazard, some of the new consortia are extremely powerful. Those who foresaw the coming of deregulation, such as John Barkshire of Mercantile House, Gordon Pepper of Greenwell & Co and Michael Hawkes of Kleinwort Benson, should have been at a competitive advantage. However, Kleinwort Benson is reputed to have lost Wedd Durlacher in a bidding competition with Barclays Merchant Bank, and Mercantile House chose to buy a smaller stockbroker, Laing & Cruickshank, rather than one of the market leaders. The foresight of Greenwell's senior partner was endorsed by the full backing of the Midland Board when Samuel Montagu announced its proposed acquisition of Greenwell & Co, although previously the Midland Bank Trust Department had been in inconclusive talks with stockbrokers Simon & Coates.

Others chose to act only when the Parkinson–Goodison agreement had been announced. According to a television programme describing the merits of synergy of the new Mercury International group, the origins of the group go back to a discussion in an aeroplane between Rowe & Pitman's partners, Peter Wilmot-Sitwell and David Brooke. As soon as they landed, Peter Wilmot-Sitwell called S. G. Warburg & Co. Lord Camoys of Barclays de Zoete Wedd was on summer holiday in Europe when he was informed of the Parkinson–Goodison agreement and immediately telephoned London to draw up a preliminary plan for a major Barclays consortium.

Three aproaches

There were essentially three approaches adopted by the UK financial institutions. The first was the direct 'no expense spared' total and immediate integration approach. Mercury International, through S. G. Warburg & Co, snapped up not only top stockbrokers Rowe & Pitman, but also Akroyd & Smithers, arguably the best firm of gilt-edged jobbers. In addition, they acquired the former government broker, Mullens & Co. Barclays Merchant Bank won the jobbing firm of Wedd Durlacher in a bidding contest, and then bought the stockbroking company de Zoete & Bevan, highly

The origins of the Mercury International group may stem from an aeroplane discussion between David Brooke (above) and Peter Wilmot-Sitwell (page 115), partners of Rowe & Pitman

Those who had adopted a 'hands-off' view were convinced that the 'no expenses spared' groups had paid too much for too little

regarded for both its gilts and generalist expertise. Inevitably, there were comments that Mercury International and Barclays Merchant Bank spent too much money too quickly. It will not be known for several years whether the strategy of Mercury and Barclays was right. What is clear is that these two groups believe in the fully integrated philosophy of combining bank, broking and jobbing skills within a single entity. The purchase of Akroyd & Smithers and Wedd Durlacher, the two dominant jobbing firms, effectively prevented any other house from pursuing the same expansion policy.

The second, the 'middle of the road' approach, implied a more cautious, and generally less expensive, strategy which still prepared the group firmly for the deregulated markets ahead. National Westminster Bank's subsidiary County Bank bought stockbrokers Fielding Newson-Smith and jobbers Bisgood Bishop. While both were highly reputable firms, neither ranked in the first tier. To strengthen its presence, County recruited extensively from outside for individual specialists. Morgan Grenfell followed a similar route, buying the medium-sized jobbing firm of Pinchin Denny and stockbrokers Pember & Boyle whose principal strength lay in gilts. Having been outbid for Wedd Durlacher, a result which it may now not regret, Kleinwort Benson acquired the very small jobbing house Charlesworth, but on the stockbroking side purchased one of the largest London firms, Grieveson Grant. Baring Brothers elected not to buy an entire broking operation but, in a move which many regard as extremely shrewd, bought the highly successful and profitable Far East division of stockbrokers Henderson Crosthwaite. While Barings clearly sees huge equity potential in Japan and the Far East, its position in the gilt-edged market was reinforced with the purchase of jobbers Wilson & Watford. Guinness Peat bought the London operation of Henderson Crosthwaite and the medium-sized jobbing house White & Cheeseman.

The third approach was to do nothing at all, or to create internal specialist skills. J. Henry Schroder Wagg has created its own in-house stockbroking firm under the name of Helbert Wagg, but it was one of the first banks to withdraw its earlier application to become a primary gilt-edged dealer. Robert Fleming has an established net market making operation in UK equities with special emphasis on electrical shares. Although Flemings' decision to make net markets away from the floor of the Stock Exchange alienated some factions of the London investment community, it has provided the bank with market making experience which most of its

Sir Jeremy Morse, Chairman of Lloyds Bank. Lloyds, like Lazards, are the exception: they have bought no-one

competitors do not possess. At this time, Flemings have not bought a London stockbroker, preferring, as in the case of Baring Brothers, to build up their already very significant skills in Far Eastern securities. Hambros Bank, even after the recent management changes, has made no significant moves to become a major participant in the new equity markets, beyond holding a minority position in Strauss, Turnbull & Co, whose primary expertise is in Eurobonds. Lloyds Bank and Lazard Brothers are exceptions in that they have bought no-one. However, while Lazards appears to have taken the view that the London securities market simply does not offer sufficient profit potential, Lloyds is building up a major group from within its own personnel and by recruiting from outside. This approach is in sharp contrast to the aggressive acquisition policy of Barclays Bank, but Lloyds still intends to become a major force and argues, with some conviction, that its conservative policy may finally prove to be the best. The bank's expansion ambitions go far beyond the London securities markets as we have seen with the attempt, which very nearly succeeded, to buy Standard Chartered Bank.

In interviews with members of the three different camps, all stressed that they were satisfied that their approach was correct. Those who had adopted a 'hands-off' view were convinced that the 'no expense spared' groups had paid too much for too little. The middle of the road sector were seen by the others as being indecisive. The most aggressive consortia saw themselves as the most far-sighted, even if the cost of entry had been high.

The winners of this unusually provocative three-sided contest will not be known for some time. It is not possible to look simply at one of the new groupings and say that 'they have the best research or the best execution capabilities'. The equation is complicated by the human factor. The financial services industry possesses few assets except for its employees. If key employees leave, the net worth of the firm may decline proportionately, unless the firm in question immediately goes out and recruits replacements from another financial institution. The game of personnel 'musical chairs' will not end after deregulation. Indeed, it may even accelerate as the gaps become more evident and customers' loyalty is put to the test. The best research and the best execution capabilities can usually be bought, albeit at a high price. There is no absolute proof that business automatically follows individual analysts or teams of analysts, but it must be a factor when the analyst is ranked number one in a particular industry sector.

Taking positions

The groups which are most likely to be successful in the deregulated equity market will be those which can instill a common corporate philosophy in the organisation's personnel, and which can offer a much broader range of services than simply skills in UK securities. Those with an essentially domestic business are likely to be at a major disadvantage. If the UK market enters a phase of sustained decline, the sector will be substantially overbroked. One-stop bond shopping will become one-stop equity shopping, which means that London houses must have expertise in North American, European and Far Eastern shares. This build-up may be gradual, but it should be part of any long-term forward planning.

Outsiders consider Mercury International to possess formidable

Mercury International's Peter Wilmot-Sitwell

The earlier reference to the probable access to the parent's balance sheet supports the case of the Greenwell–Montagu grouping, particularly as Greenwell brings substantial strengths in both equity and gilts research. The top-ranked UK research broking house, James Capel, now part of the Hong Kong and Shanghai Banking Corporation, is considered by some to have made an error in its decision to rely on agency commission business rather than making a fuller commitment to market making. 'It will take them less than a year to change their minds and in the meantime we will all be trying to poach their analysts', said a US investment banker, whose firm is a leading trader in American Depository Receipts (ADRs). Others disagree, pointing towards all-round skills as well as having the quality of management to overcome the culture differences between merchant bankers, stockbrokers and jobbers. Certainly, initial staff turnover has been minimal. Barclays de Zoete Wedd, on the other hand, was plagued with a series of early defections from Wedd Durlacher to Kleinwort Benson and, subsequently, additional Wedd Durlacher partners either left or took early retirement. Nevertheless, few doubt BZW's basic strengths over the long term. 'With Wedd's broking experience, de Zoete's strength in fixed interest and Barclay's balance sheet behind them, they cannot fail', said a rival merchant banker. Although, a London stockbroker commented, 'BZW does have some weaknesses, it ranks nowhere in the Eurobond market, it is not a power in corporate finance and has only limited coverage in international equities'.

Lord Camoys, Chief Executive of Barclays de Zoete Wedd, drew up plans for the Barclays consortium in the summer of 1983

Christopher Reeves, Deputy Chairman and Group Chief Executive of Morgan Grenfell, bought jobbers Pinchin Denny and brokers Pember & Boyle

The new UK groupings are aware that the main corporate finance competition will be in the field of mergers and acquisition

Capel's major lead in the research ratings and the necessity of all major UK institutions to have ready access to Capel's star analysts.

Citicorp's dual acquisition of Scrimgeour Kemp-Gee and Vickers da Costa draws considerable praise because of the balance provided by Scrimgeour's domestic coverage and the Far Eastern specialisation of Vickers da Costa. Scrimgeour, which is generally regarded as the number two UK research house, will be making net markets in the major companies followed by its research department, in contrast with James Capel's policy. With Cazenove placing only medium emphasis on research, Capel's stance is courageous, as virtually all its principal competitors have made plans to move relatively quickly to selective net market making.

The UK financial groups, with a major presence in corporate finance, will possibly be in a position to help their broking subsidiaries, although this association inevitably results in arguments over the effectiveness of Chinese walls and the extent to which corporate finance should be permitted to overlap with broking and net trading. A financial group with a long-established corporate finance association with an industrial client would clearly prefer to act also in the role of the company's broker. The company may find itself on the horns of a dilemma. Many relationships go back for over 50 years. The favoured broker may have moved to an unfavoured bank. A favoured bank may have acquired a broker who produced bearish reports on the company's prospects which were generally proved incorrect. If the Chinese walls remain totally intact, the corporate finance group may do little more than effect an introduction to the broker and leave the broker to prove its own worth in terms of service.

In terms of corporate finance itself, the new UK groupings are already aware that the main competition will be in the field of mergers and acquisitions (M&A). In the United States M&A business is one of the most lucrative revenue producers for the investment banks. The fee scales may be lower in the United Kingdom, but the sheer volume of M&A business in the United Kingdom makes it a natural target for the American professionals. There is considerable variation of opinions among the British consortia with regard to the likelihood of US houses achieving major market penetration. 'Any foreigners will find it difficult', said BZW's Lord Camoys, although he described Citicorp as being among the more aggressive. 'There will be particular opportunities for US houses in leveraged buyouts', said Schroder's Chairman Win Bischoff. Credit Suisse First Boston's Richard Bristow was more reserved toward the foreigner's prospects, saying that 'domestic to domestic M&A is likely to prevail for quite some time'. Kleinwort Benson's senior corporate finance executive Timothy Barker took a more realistic view when he was recently quoted as saying 'the Americans haven't come all this way not to do business'.

Win Bischoff, Chairman of Schroder Wagg, who has created its own in-house stockbroking firm, Helbert Wagg

The make-up of the new institutions

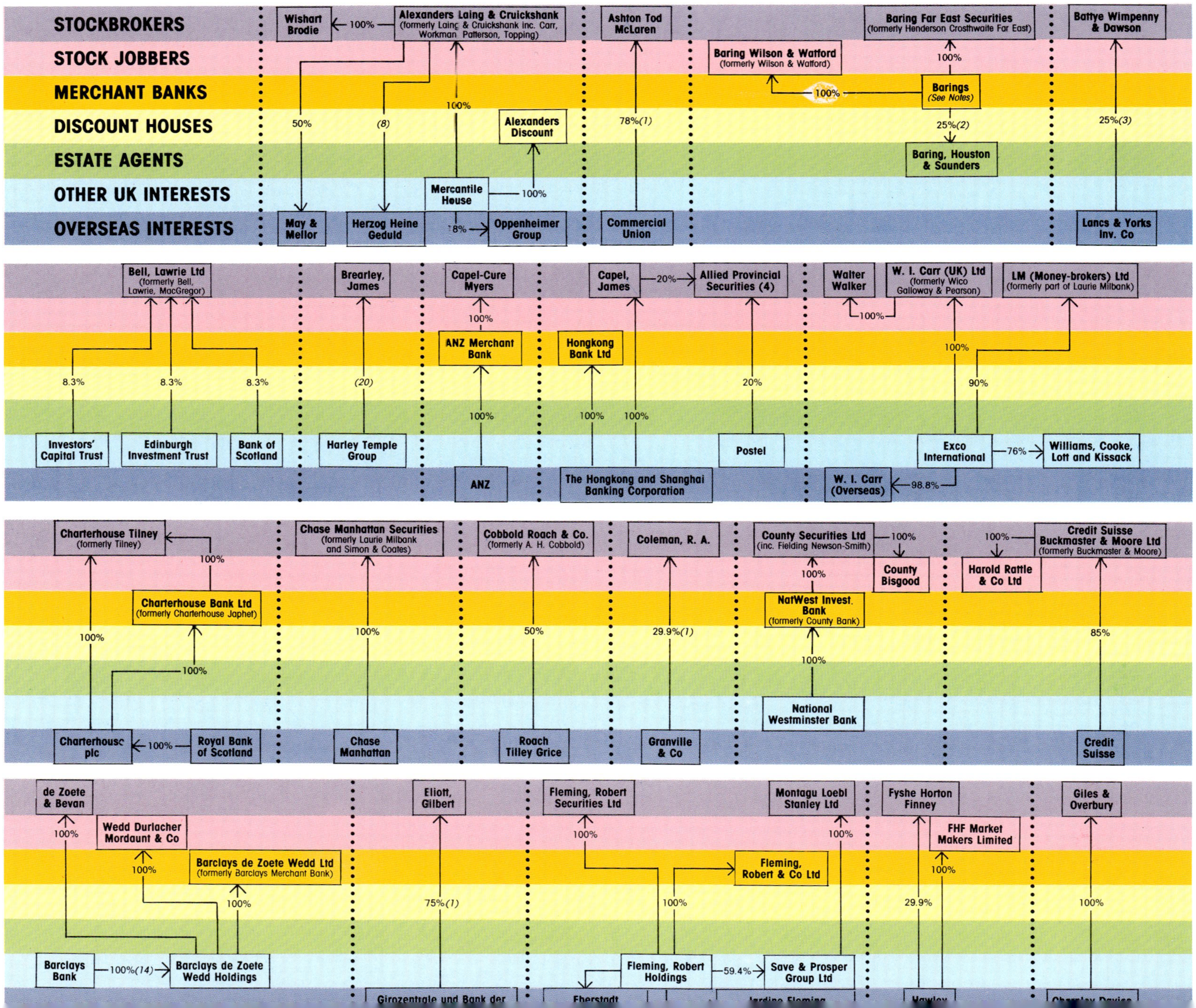

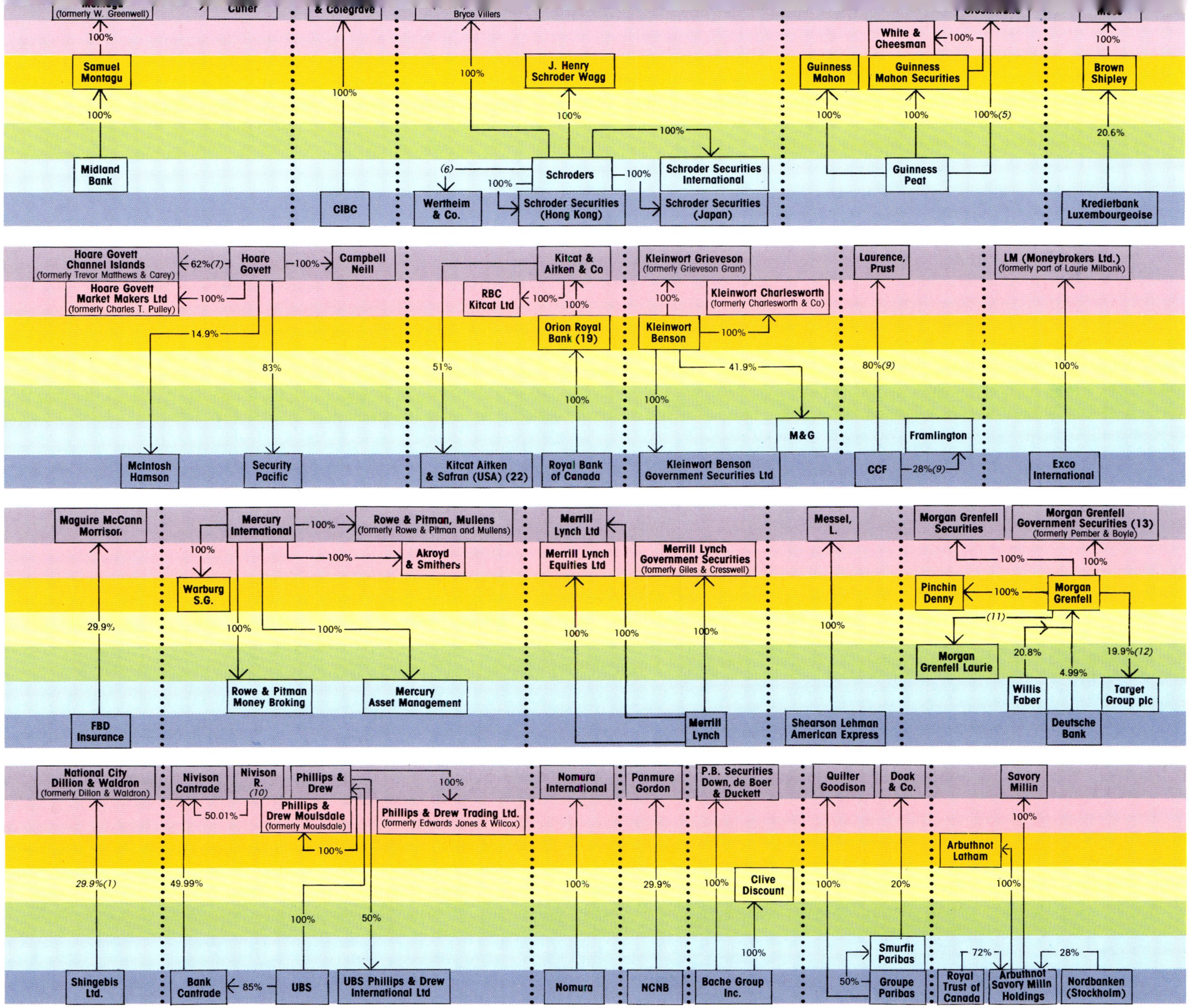

continued overleaf

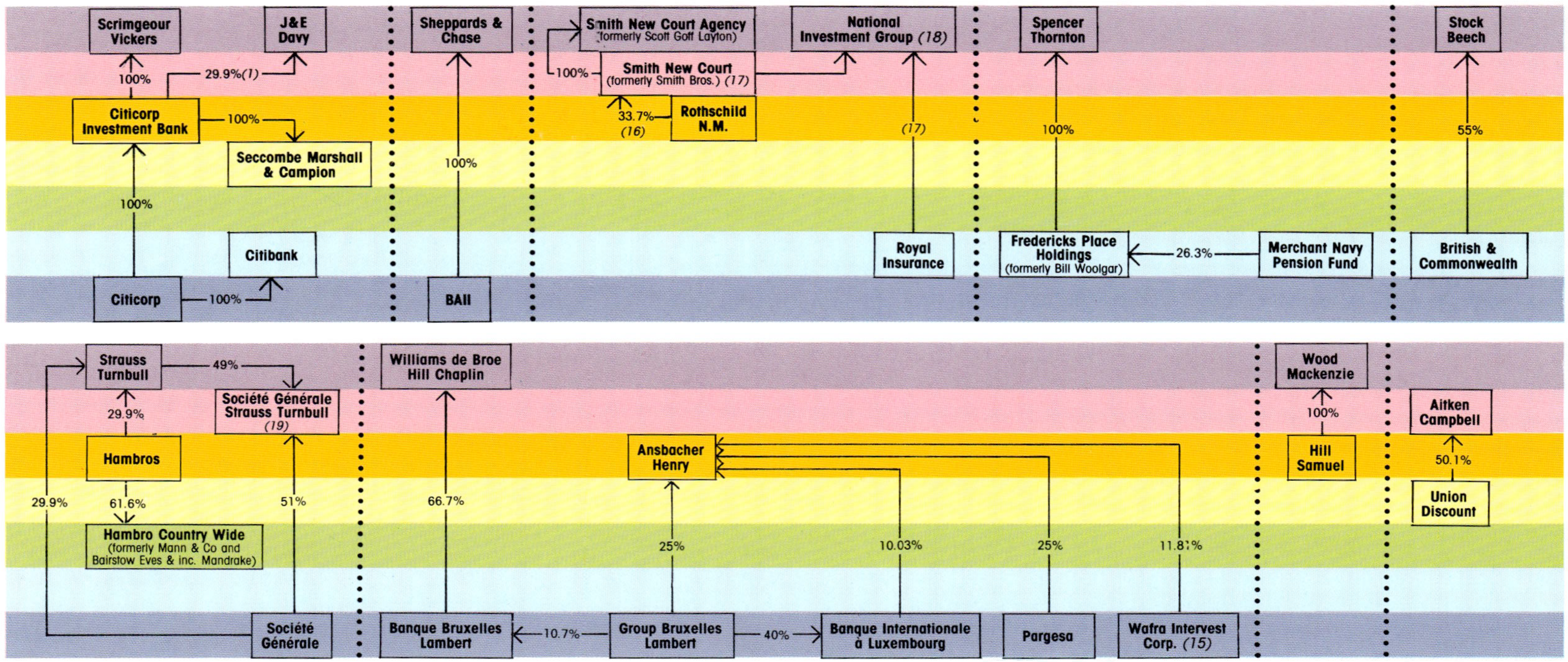

General Notes:

(a) Not all international connections have been included, e.g. Baring Investment Management Holdings Limited/Endowment Management & Research Corporation.

(b) Not all domestic connections have been included, e.g. Baring Investment Management Holdings Limited/Baring Fund Managers Limited.

Specific Notes:

(1) Buyer has announced plans to increase holding to 100% at an unspecified date.

(2) Buyer plans to increase holding to 50% when R.I.C.S. rules permit.

(3) Lancashire and Yorkshire has an option to acquire a further 4.9% of Battye Wimpenny Dawson.

(4) Allied Provincial Securities is a consortium comprising the following eight regional stockbrokers:
Parsons & Co.; Illingworth & Henriques; Stancliffe (inc. John S. Smith & Co.); Murray & Co.; Laws & Co.; John S. Smith & Co.; Westlake & Co.; William Chapman, Trease.
These companies will continue to trade under their own names.

(5) Full ownership will not take effect until October 1986.

(6) It is intended that Schroders will take a 50% holding in Weithe m at an unspecified date.

(7) Hoare Govett has an option to increase its stake to 100%.

(8) Laing & Cruickshank has formed a correspondent relationship with Herzog Heine Geduld, a U.S. OTC market maker.

(9) CCF has acquired an 80% stake in Laurence Prust's institutional equity, research and corporate finance businesses. This section continues to be known as Laurence Prust. Laurence Prust's discretionary fund management business will be acquired by Framlington and CCF has acquired a 28% stake in Framlington, Laurence Prust's advisory private client business has been renamed Laurence Keen and will remain an independent partnership.

(10) In October 1986 R. Nivison will become a part of Nivison Cantrade.

(11) A subsidiary of Morgan Grenfell & Co. Limited.

(12) Morgan Grenfell has an option to acquire up to a further 15% of Target Group PLC upon listing.

(13) Morgan Grenfell Government Securities will continue to trade as Pember & Boyle until October 27th, 1986, after which the name Pember & Boyle will no longer be used.

(14) Barclays Bank owns the whole of the ordinary share capital of Barclays de Zoete Wedd Holdings. The former partners of Wedd Durlacher and de Zoete & Bevan has an interest in BdeZWH through a holding of Founders' shares carrying, *inter alia,* the right to 14% of the votes in general meeting.

(15) Wafra Intervest Corp. is the main partner in a group of Kuwati investors which has bought a stake in Ansbacher.

(16) N.M. Rothschild holds convertible loan stock and warrants which would take its 27.4% undiluted holding in Smith New Court to 33.3% on a fully-diluted basis.

(17) Size of holding not disclosed.

(18) National Investment Group is a consortium of the following seven regional stockbrokers:
Godfray Derby (incorporating Jolliffe Flint & Cross); Hanson & Co.; Hillman Catford Board; Lyddon & Co.; Margetts and Addenbrooke; Milton Mortimer; Richardson, Chubb, Love, Rogers.
These companies will continue to trade under their own names.

(19) Eurobond market maker.

(20) James Brearly remains a partnership. The Harley Temple Group is a limited partner within the partnership.

(21) Mr Tan Sri Khoo Teck Puat owns a 27% stake in Exco International.

(22) Member of National Association of Securities Dealers (NASD).

Source: Baring Brothers

CHAPTER TEN

The strategies of the North Americans

The almost arrogant confidence of the US houses in London is probably due to the fact that they entered the one sector open to them, the huge Euromarkets, and promptly demolished the resident London opposition. Only S. G. Warburg & Co provided any real opposition, but even they could not extend their primary market strengths into the secondary area. If Credit Suisse First Boston was included in the North American category because of the interest retained by First Boston Corporation, the whitewash would be complete.

The merchant bankers' aloofness

Having established an unassailable lead over the UK houses in the free Euromarkets, it is clear that the Americans have no fear of their London competitors. Salomon Brothers' Charles McVeigh was not engaging in idle speculation when he said that 'Salomon intends to dominate the London securities markets'. He was only envisaging an extension of Salomon's dominance of several major markets in New York. This view could be disputed by a number of houses, including Morgan Grenfell Securities, whose Managing Director John Holmes was recently quoted as saying 'But in five years Morgan Grenfell will be the number one securities house in the United Kingdom'. A leading fund manager asked to judge these two equal aspirations said, 'The odds on Morgan Grenfell wouldn't be a bargain even if you were offered 100–1'.

The supreme confidence of the American houses is shown by the fact that of the major US investment banks, only Shearson Lehman bought a London stockbroker, the medium-sized firm of L. Messel & Co. The acquisition of Messel provided a sound corporate client list, a ready-made research team, good institutional contacts and specialised bond dealing skills. In addition, the London equity market is only part of the Shearson Lehman group's expansion plans abroad, where it intends eventually to become a member of all the major European stock exchanges. 'We have already participated in many substantial equity issues in Continental Europe and will expand our research coverage of these markets', said Shearson Lehman's Jacques Gelardin, whose company recently launched a fund specialising in Scandinavian shares.

The remaining major US investment banking houses displayed an aloofness bordering on disdain towards the scramble to buy UK stockbrokers and jobbers. The central theme was that the best results were achieved by 'growing from within'. This may also be interpreted as a polite way of saying that they simply did not need to buy wholesale UK firms. Archibald Cox Jnr, Managing Director of Morgan Stanley International in London, pointed out that 'our attitude has always been to expand rather than to acquire', and that 'Morgan Stanley had only bought two small specialist companies in its entire history'. Morgan Guaranty's Andrew Threadgold, a Managing Director in London, admitted that discussions with a UK stockbroker had taken place, but that 'it was not the Morgan style to acquire'. He also emphasised that purchases did not always produce the expected results in that there was no guarantee of

Morgan Guaranty's London Managing Director Andrew Threadgold: 'It is not the Morgan style to acquire'

retaining the best people after acquisition. Morgan Guaranty preferred a 'rifle approach' compared with Citicorp's 'shot-gun' policy. At Salomon Brothers International in London, Managing Directors Charles McVeigh and Sheldon Prentice said that they had also considered buying a UK stockbroking house, but 'mainly from a structural angle'. The Salomon philosophy, too, was 'organic growth from within'. Pru-Bache's Managing Director Bill Hulton said, 'We looked for a more efficient approach to the new equity market'. This was achieved by hiring three senior specialists from James Capel, who began their own, now significantly expanded, stockbroking firm within Pru-Bache. Goldman Sachs & Co, perhaps the largest equity traders in the United States, bought no UK stockbroker or jobbers but did recruit top individual specialists, including the heads of Simon & Coates' economics team, to support Goldman's new operations in the gilt-edged market. Merrill Lynch, apart from the very small purchase of gilt jobbers Giles & Cresswell, has made no major acquisitions and points to the fact that this was not necessary as it already had an established presence in both the UK gilts and equity markets.

The commercial banks

It should have come as no surprise that, while the US investment banks mainly adopted a 'hands-off' approach, the US commercial banks, with the exception of Morgan Guaranty, moved quickly into the UK stockbroking arena. In the domestic market, until the Glass-Steagall Act is finally dismantled, they are not legally permitted to make markets in equities, although they are generally all primary dealers in US government bonds. Citicorp, perhaps globally the most aggressive of all the US commercial banks, bought Scrimgeour Kemp-Gee and Vickers da Costa, two highly regarded stockbrokers. Scrimgeour Kemp-Gee is recognised as one of the top UK research houses, and Vickers da Costa is arguably the leading London specialist in Japanese and Far Eastern shares. It is too early to see if there is any natural synergy between the two, but no doubt Citicorp will eventually attempt to marry Scrimgeour's research expertise into Vickers' Far Eastern client base. Outsiders are impressed by Citicorp's astuteness in acquiring the two firms. 'Citicorp is now at the forefront of both the London and Far East equity markets for what is to them a handful of loose change', commented a US investment banker. If Citicorp is weak in any area, it is thought to be in actual trading capabilities, a view that is held of several US commercial banks. But on closer examination this is not a very accurate assessment. Citicorp is a primary US government bond dealer, and the new gilt-edged market is unlikely to cause it any concern, particularly as the UK gilt-edged market is only a fraction of the size of the US government bond market. Citicorp acquired direct gilts expertise through the purchase of Seccombe, Marshall & Campion. In addition, Citicorp, in line with most other leading US commercial banks, has been making net markets in Eurobonds, Euronotes and other short-dated paper for years. Trading skills are in fact an integral part of the operations of all US commercial banks.

Chase Manhattan also purchased two London stockbrokers, Laurie Milbank and Simon & Coates. Both are in the second-tier category, although Laurie Milbank brought Chase Manhattan some gilt expertise, and Simon & Coates were in the upper ranking of UK research. Chase Manhattan's resolve to become a leader in the UK securities industry is not doubted, and the bank has already committed more capital than most of its competitors. Nevertheless, the lead-up to deregulation has not been easy for Chase Manhattan.

Sir Peter Tapsell of James Capel, now part of Hong Kong and Shanghai Banking Corporation

There have been a number of well publicised defections, including whole teams of professionals

There have been a number of well publicised defections, including whole teams of professionals. While this is unavoidable in a market short of specialist skills, Chase has been affected more than most. 'With their capital they should have bought bigger and better, but if they can overcome the problem of staff haemorrhage, Chase will be an important factor', said a Managing Director of a rival US commercial bank. Security Pacific's very early purchase of stockbrokers Hoare Govett is widely envied. 'They bought top quality for what proved to be bottom dollar', commented a senior London broker.

Michael Pragg of Simon & Coates, which with Laurie Milbank was bought by Chase Manhattan

With Bankers Trust electing to concentrate on the gilt-edged market through its BT Gilts subsidiary, the one name missing from the new London participants is Bank of America. The domestic problems of the bank are well known, but Bank of America certainly intended to participate in the new market, and if its fortunes improve at home it may well return at a later stage. It was one of the original 31 candidates approved to become primary gilt-edged dealers, but withdrew at an early stage. Perhaps in line with its diminishing importance in the United States, the bank has been plagued with staff defections, although outsiders continue to rate its London operations highly. A possible point of re-entry to the London securities market would be through a purchase which could be linked to its highly successful discount broking subsidiary Charles Schwab, which has no doubt examined the possibilities of establishing a no-frills discount service in the United Kingdom.

Of the Canadian banks, only Royal Bank of Canada has decided to make a major commitment in the gilt sector through its subsidiary Orion Royal Bank, which will be one of the 27 primary dealers. Canadian Imperial Bank of Commerce, through its London subsidiary CIBC Ltd, has purchased Grenfell & Colegrave, a small- to medium-sized London stockbroker. Despite its commitment in the highly competitive gilt-edged market, Orion Royal also chose to purchase only a medium-sized stockbroker, Kitcat & Aitken, in which Jacob Rothschild, through Rothschild Investment Trust, had once owned a minority stake. Both Orion Royal Bank and CIBC Ltd have built up enviable niches in the Eurobond market during the past, and will be looking for similar opportunities in the deregulated London market.

'What the Yanks want out of the City'

A recent article was entitled 'What the Yanks want out of the City of London'. The answer, unfortunately, was never quite provided. In reality, anyone who has had the opportunity of transacting business with the US houses over a period of years will be able to identify their objectives with some precision. There is no doubt whatsoever that they intend to become the most important force in the distribution and trading of UK securities. The US houses believe that they have stolen the march on the Japanese houses, who they think will pose a real threat once the Japanese make their major moves to enter the London market. The US houses also intend to become an important force in UK domestic corporate finance and to introduce a whole range of new securities based on instruments which have been successfully marketed in the United States.

Are these objectives overly ambitious? They may not be as far-fetched as they appear. The confidence of the US houses stems to a large extent from their success. They are extremely profitable as well as successful. This does not imply any disrespect towards their UK counterparts. The majority of US investment banks stayed on the sidelines while the bidding war for UK stockbrokers and jobbers was in progress, not because they did not admire many of the firms involved, but because they thought that the asking prices were too high. This may not always have been the right decision. Shearson Lehman is very pleased with its purchase of L. Messel, and Citicorp's purchases of Scrimgeour and Vickers da Costa are widely considered to have been a coup. Security Pacific's initial purchase of a minority stake in Hoare Govett not only showed other US banks the way, but underlined the fact that values then were a fraction of the levels today.

It may, nevertheless, be true that some US investment banks never seriously entertained the purchase of a UK stockbroker or jobber. The philosophy of 'growth from within' has many supporters on Wall Street. As US houses looked at the London market, they thought that they already possessed most of the skills within their own firms to take a reasonable share of the market. The missing ingredients were essentially research and UK corporate finance expertise, which could be bought. Far-sighted US investment banking houses such as Goldman Sachs had already been building up their own in-house UK corporate finance expertise for more than five years. The US houses had no scruples about raiding UK houses for top talent: on Wall Street high-level personnel raiding has become a time-honoured profession. Nor were they inhibited by salary scales: on average an American moving from New York to London was paid three or five times as much as his UK counterpart. It was a persuasive influence which triggered huge salary rises in the City.

The gilt-edged market

It seems clear that the first targets of the US houses were the London gilt-edged market and the London equity market. The overall policy towards the gilt-edged market was simplistic: all that

Will the Stock Exchange floor be needed any more?

Whether this proves to be akin to a Polish cavalry troop fighting a German Panzer division is essentially immaterial

was required was to transfer US trading strategies and patterns to the UK gilt market. Of course, refinements would have to be made, but the intended conclusion was already apparent. US houses would dominate the UK gilt-edged market. The opposition would be Mercury International, with its Akroyd & Smithers subsidiary, and Barclays de Zoete Wedd, who own the former Wedd Durlacher jobbing firm. As Akroyd & Smithers and Wedd Durlacher had dominated trading in UK government securities, it was imperative for the US houses to impose their over-the-counter skills onto what had been a floor-based system. The fact that Merrill Lynch, through its subsidiary Giles & Cresswell, 'bought' almost 20 per cent of the UK gilt market was a harbinger of things to come. Many UK houses scoffed at Merrill Lynch's initial market share, but all would have liked to emulate the achievement. Merrill Lynch proved beyond any doubt that no domestic territory is sacrosanct. The intention of the major US trading houses is to acquire effective control of the UK gilt-edged market within a period of two or three years. US houses see themselves as better capitalised and better attuned to the day-to-day problems of accepting risk. A £50 million trading loss could be absorbed without recourse to the controlling shareholder. How would UK houses accept a loss of £50 million, which might be incurred within a matter of days? That is the key consideration in trying to assess the perception of risk-taking in the decade ahead. 'Wedd and Akroyd couldn't write off the equivalent of £50 million in the days of single capacity without having to ask for additional financing', said one US investment banker.

If the US houses have the advantage in terms of capital and the ability to accept much higher margins of risk, can they compete against the research of the UK gilt-edged specialists? Research may be valuable, but is it essential to the new international fund manager whose decisions are based on global considerations? Perhaps the most unrecognised aspects of the US houses' entry into the gilts market is that they do not have to make an immediate profit. Few UK houses can afford that luxury. US houses can offset losses in the United Kingdom against profits in the Japanese government bond market, hedged programmes and underwriting participation in other capital markets. The mistake some commentators have made is that they genuinely believe losses in the gilt-edged market are unexpected. They are not. The most essential factor is, who can best absorb the losses? Perhaps there will be no losses at all, but that would assume a major bull market. The real competition in the new London gilt market is between the City's establishment and the newcomers. Whether this proves to be akin to a Polish cavalry troop fighting a German Panzer division is essentially immaterial. Within a period of two years, it appears, the majority of UK gilt-edged business will be controlled by foreign-owned companies. The great era of Akroyd and Wedd Durlacher's monopoly has gone. In its place will be a new age where the gilts market's domestic identity will be replaced by an international market, and its securities will be traded in a way that will break all former traditions. In the final event, it will probably be good for the market. The government must encourage wider foreign ownership of sterling securities. It may be of little concern to the government that they are distributed by non-UK houses.

Equities

The policy of the North American financial groups towards the London equity market is likely to be less aggressive than towards gilts. As only a few have bought London stockbrokers, they do not have the research teams in place to support a full across-the-board trading operation. In these circumstances the US houses are likely to concentrate on the UK top 100 companies while they build up their research departments. They are already familiar with the largest UK companies through trading of the shares in ADR form from New York. The emphasis will be on block trades at which the North American houses excel. This may be good news for those institutions who conduct a large amount of research in-house. Net markets will be available in all the top UK 100 shares, dealing margins will be wafer-thin and block trades will be in sizes never before seen in the City. The US commercial banks do have less expertise in net market making than their investment bank counterparts, but this can be overcome by selective recruiting. The large US investment banks argue that a good block trader can trade anything, irrespective of its nationality. Over a period of time, however, a nucleus will be formed in equities which will be based on the traditional US concept of linking trading sales and research. As this progresses, North American houses will widen their coverage of UK shares beyond the top 100. Up to now the US houses have shown only limited interest in retail clients, with the exception of 'very high net worth' individuals. Even Merrill Lynch and Prudential-Bache, who have traditionally courted the small investor in the United States, are reputed not to be soliciting business from

Only the latest US technology will be imported

any retail customers with a cash or cash equivalent net worth below £100,000. Judging by interviews, the interest of US financial groupings in retail custom, beta or gamma stocks or USM securities would seem to be limited.

Corporate finance

The US attitude towards corporate finance in the United Kingdom is that patience will bring its own rewards. They may intend to dominate the gilt-edged market within a few years, but they are under no illusions about similar ambitions in corporate finance. The position of banks such as Morgan Grenfell, Kleinwort Benson and S. G. Warburg is not totally unassailable, but it will be difficult to make any serious inroads for some time. The acknowledged UK lender among the US investment banks is Goldman Sachs & Co, who formed their own London-based corporate finance team several years ago. In 1985 Goldman Sachs completed 27 M&A transactions, the majority of which were uncontested bids. More recently Goldman Sachs recommended blocking tactics in the United States, which effectively prevented Lloyds Bank from buying any more shares in Standard Chartered Bank in the open market. This probably saved Goldman Sachs's client Standard Chartered from being taken over by Lloyds Bank.

Up to now US investment banks have generally been only one of a group of bankers acting as advisers in a UK domestic takeover bid or defence. They fully intend, however, to play a leading role within the medium term. Goldman Sachs would appear the most likely to achieve this position, but some UK bankers rate the chances of Citicorp and Morgan Guaranty equally highly. They point out that both Citicorp and Morgan Guaranty can offer UK domestic clients a full range of commercial and investment banking services and, in addition, they have virtually unlimited capital resources. This could be extremely important for soliciting new corporate finance business in the event of a major takeover bid, where these two large US commercial banks might be prepared to finance the whole offer. This occurred in the recent offer by the Australian industrial corporation Elders IXL for Allied Lyons, the UK food and brewing combine. The multibillion pound offer was financed by a group of banks led by Citicorp.

Corporations value their associations with banks who have been prepared to finance their public acquisitions. In the case of the recent contested bid by Lloyds Bank for Standard Chartered Bank, it should have come as no surprise that one of Standard Chartered's White Knights was Australian businessman Robert Holmes à Court, whose fast-expanding Bell Communications group had been regularly financed by the Standard Chartered group.

A dark horse in the US investment and commercial banks' efforts to become global corporate finance forces is Drexel Burnham Lambert. Whatever some professionals think about acquisitions financed by the issue of high-yield or junk bond offerings, no-one can dispute Drexel's success. Initially, Drexel simply specialised in the distribution of high-yield bonds, but later it realised that the sales network had expanded to such an extent that newly-created junk bonds could be a self-financing mechanism for corporate predators. Drexel has subsequently financed a series of multibillion dollar takeovers, mostly hostile but mainly successful, and has prospered to such an extent that the privately-owned firm's capital is now in the region of US$1 billion, a level which comfortably exceeds that of most UK merchant banks. Will Drexel's junk bond financing concepts be as successful in Europe as they have been in the United States? Drexel has high hopes, but there may be some opposition from the regulatory authorities who would probably not wish to see low-quality paper being widely sold to the public. In addition, Drexel's corporate finance effort in Europe is virtually unknown compared to the highly visible US groups, and the Takeover Panel might be less than enthusiastic about bids which include very little cash, and where the backing for the junk bonds is often the assets of the target company.

CHAPTER ELEVEN

The European and the Japanese groups

It came as some surprise when the original 31 primary gilt market dealers were announced that only one European name, Union Bank of Switzerland, was included. Admittedly, CSFB (Gilts) was naturally on the list but was CSFB either British, Swiss or American? Taking into account the majority shareholding of Crédit Suisse, it should almost certainly be classified as a European house, but there were still some surprising abstensions. If Union Bank of Switzerland, through its subsidiary Phillips & Drew, was participating, where was Swiss Bank Corporation, the last of the 'Big Three' in Switzerland? Deutsche Bank had purchased a minority interest in the UK merchant bank Morgan Grenfell, but it was not participating directly itself, and neither did any of the other large German banks. The French were nowhere to be seen in the new gilt market, although several subsequently purchased UK stockbroking firms.

The Japanese institutions

The absence of the Japanese seemed difficult to understand when Nomura Securities was on the verge of becoming a member of the London Stock Exchange, and when the collective influence of the four largest Japanese securities houses was sufficient to demand a major position in any of the world's principal capital markets. The answer to the apparent Japanese reticence is probably that they are not being reticent at all. Instead, they will adopt their normal 'wait and see' attitude, which has characterised their approach to the Euromarkets. In the public Euro debt market, the Japanese houses were for a long time content to play a junior role to the traditional primary market power-houses such as Credit Suisse First Boston and Deutsche Bank. By 1986, however, all of the four major Japanese securities houses were among the top dozen and the leader, Nomura Securities International, ranked in the first position as measured by the number of individual issues managed, although only third if measured by total volume.

In the London markets, the Japanese houses have elected not to buy any stockbroking firms, but this appears to have been a deliberate policy and may disguise a 'grand design policy for the future'. Speculation that Kleinwort Benson was a bid target for Nomura may have been no more than idle market gossip, but no one doubts that the 'Big Four' Japanese houses have the financial strength to acquire almost any financial institutions that they set their sights upon. In terms of size, Nomura Securities even outranks Merrill Lynch, and both have capital exceeding US$2 billion. The Japanese houses are also extremely profitable. In 1985 Nomura Securities reported profits of US$850 million compared to US$557 million for Salomon Brothers.

At a later stage, the wholesale purchase of a British financial institution might make more sense to the Japanese houses than creating their own trading and sales network. In the years after deregulation of the London markets it will soon become evident that some participants are weaker than others. If, as seems likely, capital will be one of the ingredients for survival, the Japanese houses could be in an ideal position to purchase an established name at a discount price.

The Japanese banks, as opposed to securities houses, may also prefer to enter the UK securities markets via acquisition. They are already a formidable force in the Euromarkets and, as can be seen from their expansion in the Swiss franc sector, they are keen to establish a management role. The Swiss subsidiaries of Japanese banks are already lead managing new issues, a progression which has been assisted by the fact that a high proportion of new Euro Swiss franc issues are for Japanese borrowers.

As Japanese foreign investment capital flows, which in 1985 amounted to more than US$50 billion, are essentially directed towards bonds rather than equities, it would seem logical to assume that the first Japanese move will be in gilts. If this is correct, the argument for buying an integrated primary market dealer, complete with inbuilt trading and sales teams, would be even more convincing. To date Japanese buying of gilts has been spasmodic but this will change. For Japanese institutional buyers the most important considerations are liquidity and a broad selection of maturities. The majority of Japanese international investment is therefore currently being channelled into US Treasury bonds, despite the persistent weakness of the US dollar, which has totally negated the premium yield differential of US dollar bonds over domestic yen issues. As the new gilts market will be among the most liquid in the world, it could attract more Japanese institutional buying interest unless the Japanese domestic purchasers take a negative view of sterling.

UK equities, on the other hand, will hold only limited appeal for Japanese buyers. Foreign purchases of overseas equities, even in terms of total aggregate, are minimal in relation to investments in bonds. Any move to diversify domestic portfolios would probably begin with US equities, with which Japanese fund managers are more familiar. As part of a much longer-term approach, the Japanese securities houses are building up a global research operation. This will necessitate membership on all leading stock exchanges, but in itself this does not signify a major change of Japanese investment philosophy away from domestic securities. If the Japanese houses wish to purchase an established overseas client base, this can be most easily achieved by direct acquisition of an overseas bank or financial institution, such as Sumitomo Bank's acquisition of the Lugano-based Banca del Gottardo.

The European institutions

The most significant European purchase of a London stockbroker was Union Bank of Switzerland's acquisition of Phillips & Drew, one of the largest and most highly regarded firms in the City. In order to strengthen its market making capabilities, Phillips & Drew acquired a small gilt-edged dealer, Moulsdale Rensburg, and the equity jobbing firm of Edwards, Jones & Wilcox. However, even with the capital backing of Union Bank it is unlikely that Phillips & Drew will rank higher than the middle levels of Division Two as a primary dealer in gilt-edged securities. While Union Bank has more funds under management than probably any other European bank, its traditional client base has historically shown little interest in UK government securities due to the persistent weakness of sterling versus the Swiss franc. The appeal of Phillips & Drew to Union Bank probably lies more in its all-round UK equity expertise and its position as one of the top three London research houses.

Compared to the estimated US$100 million which Union Bank is said to have paid for Phillips & Drew, Crédit Suisse's acquisition of Buckmaster & Moore is relatively modest, but Buckmaster has an extensive corporate client list comprising mainly small- to medium-sized companies. In order to provide trading support, Crédit Suisse also purchased the small jobbing firm of Harold Rattle. Both Buckmaster and Rattle will be integrated within the London offices of Crédit Suisse rather than Credit Suisse First Boston. After looking closely at a number of firms, CSFB has not bought any London stockbroker or jobber, although it will become a primary gilt dealer and will certainly be one of the leading houses in Euroequities.

Girozentrale und Bank der Österreichischen Sparkassen will be one of the few Austrian financial institutions represented on the London Stock Exchange, following its acquisition of brokers Gilbert Elliott, known principally for their specialisation in UK corporate debentures. Bank Cantrade, in which Union Bank of Switzerland has a majority stake, bought R. Nivison, a medium-sized but highly successful stockbroking firm. The only Belgian representation is through Banque Bruxelles Lambert's acquisition of Williams de Broe Hill Chaplain, and Banque Bruxelles Lambert has expanded further into the London capital market through buying a minority interest in the merchant bankers Henry Ansbacher.

Although Société Générale formed its Eurobond trading link with Strauss Turnbull as long ago as 1980, the other French banks have not been aggressive participants in the new London securities markets. Only Paribas has bought a major London stockbroker, Quilter Goodison, and while Crédit Commercial de France purchased the small firm of Laurence Prust, its strategy may be more towards money management, where it has bought an interest in the Framlington Group. The Paribas purchase of Quilter Goodison is intriguing because Paribas only became a buyer at a very late stage. Originally, Quilter Goodison had agreed to merge with the insurance company Skandia, but the acquisition could not be completed because of a legal technicality. The merger with a European bank would seem ideal for Quilter Goodison, who are probably the best London-based research house in Continental European equities. Paribas has been successful in strengthening its retail client base outside France and now has major operations in Switzerland, Holland, Belgium and Luxembourg. Quilter Goodison has also been at the forefront of new concepts designed to attract retail customers, such as opening branches in major stores. Paribas is a major force in the Eurobond market and will no doubt use Quilter Goodison's research and distribution capabilities for its ambitious expansion plans in Euro-equities. Because of its strong position in a number of capital markets, Paribas must be a possible contender to become a primary gilts dealer at a later stage.

THE PROSPECTS

CHAPTER TWELVE

New technology for the new markets

It was clear at the earliest stage that completely new operating systems would be required if the deregulated London markets were to function efficiently. The example had already been set in New York where, in the 1970s, Salomon Brothers had equipped a huge state of the art dealing room with the most advanced technology of the time. The only problem with state of the art is that it never stands still. Every new dealing room, dealing system or information system soon becomes yesterday's state of the art.

The basic technological requirements for the London capital markets will be to assist:

1 the execution of transactions;
2 the constant monitoring of prices being displayed by market makers;
3 the accurate and fast settlement of transactions between two parties;
4 external information and research;
5 internal information and research; and
6 improved international communications.

The number of systems available to prospective buyers provides the widest possible choice. One of the accusations that could never be made of London's new financial consortia is reluctance to spend money on improved technology systems. Some of the largest companies have each spent more than US$20 million on new installations. The main problems have been integration of existing systems and selection of the right new systems. The first has often proved to be a nightmare, which has usually resulted in having to start completely afresh. In the *Euromoney* issue of February 1986 it was pointed out that in the case of the BZW merger, Barclays Merchant Bank was using IBM equipment, de Zoete & Bevan had a computer bureau for its back office while the front offices had Hewlett-Packard installations, and the Wedd Durlacher jobbing arm of the new BZW conglomerate was based on ICL systems. Similar situations could be found elsewhere. In the Mercury International grouping, S. G. Warburg was using a Sperry mainframe, stockbroker Rowe & Pitman used ICL computers, and the gilts subsidiary Mullens & Co employed a systems bureau. These are only two examples. The result was that whole existing computer systems were made redundant overnight. In their place came newer and more advanced systems. The effectiveness of the replacements was always open to doubt. Peter Dale, a senior director of County Bank, succinctly summarised the situation when he said, 'By definition you never get more than 70 per cent of what you really want, but at least we know it works'.

Total technological compatability could never be exactly perfected within the new London capital markets because no-one knew all the final amendments to the Financial Services Bill, as well as those that would subsequently be made even when the Bill became law. Many houses simply scrapped their internal systems and looked to outside advice for the replacement. In line with the demand for traders, salespersons and compliance lawyers there came the requirement for skilled programmers and data systems personnel. External service companies had to be complemented by trained in-house computer specialists. The role of the in-house specialists was to create tailor-made systems for the new dealing rooms and the operations staff which supported the trading system.

There are only a handful of firms who will openly admit that their houses are technologically ill-equipped for the deregulation of London's markets. What is more likely is that after Big Bang several houses will discover that they have installed the wrong systems for their own personal needs within the new market-place. This should not imply incompetence or lack of forward planning; the fact is that no-one knows exactly how the new markets will operate, and systems installations have therefore had to incorporate at least a degree of guesswork. The largest houses, with their very substantial capital resources, have been able to cover almost every option. National Westminster Investment Bank, for instance, had only one operating system at the end of 1985, but will have a further 21 systems by the end of 1986. Many of these are based on totally different hardware which will complicate the facility of achieving direct interfacing. The rationale for using totally different hardware and software is that certain systems are clearly superior in specific financial areas.

The SEAQ system

The London Stock Exchange's central computer installation will be based on the SEAQ (Stock Exchange Automated Quotation) system, which has been specifically designed to support dual-capacity trading. The aims of SEAQ can essentially be divided into four general categories:

1 the collection of prices from all the accredited market makers;
2 the increased participation of non-London Stock Exchange member firms, particularly foreign financial institutions;

The number of systems available to prospective buyers provides the widest possible choice

3 access, on the broadest possible basis, to market price makers' quotations; and
4 improved standards of market codes of conduct and the assurance of 'fair' dealing prices.

The objectives of SEAQ International in these areas were succinctly described by Peter Cox, the Project Manager of International Markets for the London Stock Exchange, in a speech in November 1985 sponsored by the Stock Exchange:

Quotation collection
The SEAQ International system provides the focal point for the collection of bid and offer quotations from competing market makers. These dealing prices are then available for distribution to computer terminals on the desks of dealing firms and investors. Someone with a terminal showing these prices will be able to see an individual market maker's quotation, telephone him and expect to deal at that price.

The system currently has quotations in over 500 leading non-UK stocks and this number is increasing as more market makers participate. Most of these stocks are quoted by more than one market maker. Where there are three or more market makers in a particular stock, the quotations for that stock are brought together in a 'composite display' so that they may all be seen on the display screen simultaneously and can quickly be related one with another. The investor is, of course, looking for the best price for the security in which he is interested. The composite display presents all the alternatives to him so that he can see at a glance the best price and the market maker who is quoting it.

Participation
At the outset it was intended that SEAQ International would cater for the majority of business done in foreign stocks in London by net market makers – i.e. those buying and selling at a net price rather than on commission. Clearly this meant that the Stock Exchange International Dealerships – member firms set up specifically to deal in dual capacity in foreign stocks – would be included, but it also meant that many large foreign securities houses had to be included since they do a large part of the business. Market making firms currently outside the Stock Exchange have therefore been invited to contribute their quotations and to become members of the Exchange in due course when the rules, which change in October 1986, allow it. At the time of writing there are over 35 market makers contributing quotations to SEAQ International, with half-a-dozen additional firms waiting to be included and the members growing rapidly. Among this total are all the major US, Canadian and Japanese houses making net markets in London.

Each of the participating firms has been invited to sit on a practitioners panel which will advise the Stock Exchange on the way in which this market should be developed.

Distribution
The London International market is serving investors in the UK and many other parts of the world, especially Europe. Any communications system intended to carry these dealing prices direct to the investment manager's desk must therefore have a wide international coverage. While the Stock Exchange's own TOPIC (Teletext Output of Price Information by Computer) system has extensive coverage within the UK (and this will be expanded considerably during 1986 to cater for SEAQ in domestic stocks), we have very few terminals installed outside the country. It is clear, therefore, that TOPIC alone cannot provide the distribution mechanism required. Nor, if one is looking for even coverage everywhere in Europe, can any one network operator. For example, Reuters, which has extensive coverage of much of mainland Europe, has much lower penetration than Telekurs of the Swiss banks, who are big investors in international equities.

Our policy, consequently, is to join forces with all the commercial network operators in order to display SEAQ International prices as widely as possible. The SEAQ system, having collected the quotations from market makers, feeds them to each of the network systems, including TOPIC, on an equal footing. Thus the evident strengths of the network companies can greatly increase the reach of the Stock Exchange.

ARIEL

While SEAQ is reported to be having some teething problems, the system should help to dispel one of the principal criticisms of the London market, that it had seriously lagged behind other international financial centres. It is true that ARIEL (Automated Real-Time Investments Exchange Limited) had been founded in the early 1970s. The ARIEL concept was entirely sound, although, in retrospect, perhaps slightly ahead of its time. ARIEL provided a system through which its members could input bids and offers for shares which would be simultaneously displayed on the screens of the other subscribers. The principal advantages were, (a) anonymity – the ARIEL service was not in fact totally different from the inter-broker dealing systems of today, and (b) the lower dealing costs. Transactions were completed for a small fixed charge which

As firms have merged, computer systems costing millions of pounds have been abandoned because they were incorrectly designed

bypassed both jobbers and brokers. At the time there was considerable concern among Stock Exchange members about the advent of ARIEL. The threat of ARIEL was sufficient to instigate a Special Committee which, not surprisingly, recommended that its members should not be permitted to join. ARIEL's American counterpart INSTINET had secured only minimal market share with a similar system, and the London Stock Exchange was convinced that ARIEL would fare no better. They were proved correct. ARIEL's impact was insignificant, perhaps due to the Bank of England's decision not to trade its gilt-edged positions through the system. The more important result of ARIEL's arrival was the Stock Exchange's decision to create its own system, which immediately recorded transaction price levels and would inform institutional investors of blocks of shares available through a particular broker. This was amalgamated into the TOPIC system which began operations in May 1980.

It is difficult to see the reasons for the relative failure of ARIEL which is, in essence, a blue-print for the new markets of today. It should have exploded the myth that face-to-face dealing, as was the traditional custom on the floor of the Stock Exchange, was always in the best interests of the client. It should have alerted the London Stock Exchange that an over-the-counter Euromarket or NASDAQ (National Association of Securities Dealers Automated Quotation) system operation was the shape of the market in the future. It should have demonstrated to the London Stock Exchange that the major institutional buyers, when drawing their own investment conclusions independent of a broker's advice, were not always prepared to pay for the research which the brokers provided.

The City's failure to match the technological advances being made by its US and European counterparts, a situation now being rapidly redressed, can be traced to a number of specific factors. Prior to the deregulation of markets and the cessation of fixed commissions, the UK stockbroking business was highly fragmented. Capital limitations were an inevitable consequence of the partership system. Specialisation in certain niche areas appeared to lessen the necessity for all-round generalist expertise which could only be provided by access to a sophisticated data base. In general, London houses were risk averse compared to the US companies who thrived on controlled risks. In most instances the control was provided by the installation of tailor-made computer systems. The systems are not always foolproof. Even the most professional trading houses have discovered computer-inspired straddles that do not straddle, and computer-assisted trading prices, used increasingly on short-dated money market instruments, cannot always identify market anomalies such as a price distortion caused by a bear squeeze.

Technology for the sake of imitation or being able to keep up with the competition is a temporary solution to a longer-term problem. As firms have merged in the City, computer systems costing millions of pounds have been abandoned, not because they were obsolete, but because they were incorrectly designed for the end users in a consortium form rather than as individual subscribers. These obsolete systems are being replaced with systems costing tens of millions of pounds. It is impossible to guarantee at this stage that the money is being well spent in terms of competitive edge, better use of the internal data base, surveillance against fraud or cost savings through more efficient settlement of transactions. For the new financial consortia, management will require satisfactory systems to cover all these areas, and without doubt there will be some initial teething problems as well as the problem of installing the equipment by the deadlines set for the opening of the new market.

Approach to new systems

There is certainly no perfect approach to the installation of new computer systems, and the widely disparate attitude among the market's leading practitioners suggests an element of confusion, even taking into account that what is the ideal solution for one house, could cause chaos in another. The most sensible solution would probably be to formulate a planning strategy which incorporates the installation of systems over a period to reflect the expectation of business conditions in the future. In itself, this already constitutes a problem because many earlier predictions proved to be hopelessly inaccurate, and this often resulted in selection of equipment being based on the most defensive, and therefore probably not the most productive assumption. The forward planning time should, therefore, probably not attempt to look ahead for more than two years and accept that, in the highly erratic securities industry, this is still an optimistic horizon. Those financial institutions with existing equipment which performed efficiently in the old markets can only hope that they are as successful in the deregulated market-place. As some changes will

The most important hurdle is acceptance of the new equipment by the staff

almost certainly be necessary, management must firmly establish a list of priorities.

Apart from the efficiency of the systems themselves, the most important hurdle is acceptance of the new equipment by the staff. There is no point in providing abundant support systems if they are not used to reasonable effect. Once they are seen to provide genuine assistance to the market practitioners, they will quickly be recognised as essential. With some firms çommitting literally billions of dollars in the market-place, key staff must be able to believe that they are better prepared to manage risk. The spectrum of new financial instruments has expanded to such an extent that it is almost impossible to evaluate the full extent of the risks involved. Every house may maintain that it can efficiently contain risk through intelligent hedging techniques, but few will insist that their programs are infallible. Management must have a very clear view of the type and breadth of risks which they are prepared to assume. In the new gilt-edged market, for instance, this will be vital in minimising risks in the early stages while evaluating the potential volume of business and the level of profitability, if any, which it might provide.

In the United States, where the sophistication of computer systems is almost unrivalled, it is generally acknowledged that even the most advanced equipment is not the perfect answer to risk control. Instead, it is used as a major indicator for taking positions in securities, but the final decision as to whether tens of millions or even billions of dollars should be committed to the market-place rests essentially with the in-house trading room management, who will be assisted by a team of computer professionals. In a typical trading room, computer support is an inherent part of identifying anomalies which will lead to profitable arbitrage opportunities, a window for new issues and potential swap suggestions. If and when the world's major capital markets are effectively integrated, bond specialists will be relying very extensively upon computer support programs to assist identifying the best overall rates of return which can be achieved for themselves as a principal and for their clients.

Technology and its problems

Rather than listing the various claims of all the computer suppliers or consultancy companies, who all claim that their systems are the best, a more philosophical view towards technology and its attendant problems was recently taken by John Skerritt, a partner in charge of Arthur Anderson's consultancy practice in banking in London. In an article written in March 1986 he commented:

> On the assumption that the major institutions will be able to implement acceptable transaction processing systems for their mainstream business in time for Big Bang and other impending changes, and recognising that such systems will be fragmented, it is reasonable to ask whether it will be feasible to enhance and integrate these systems at a later stage. If not, what can be done now to mitigate this problem? Fortunately, there are a number of steps that can be taken. First, the matter of infrastructure planning should be addressed. This includes primarily the introduction of comprehensive information management and transmission systems, such as suitable local area networks to support not only text processing but, ultimately, the electronic office; international telecommunications and telex facilities linked to the local area networks; and voice mail and squawk boxes, primarily for use in dealing rooms. Those banks which recognise that they are, fundamentally, in the information business, and which position themselves to acknowledge this fact, will be best placed to prosper. The product they offer to customers is information-based and, for the most part, it cannot be delivered without automated supporting systems.
>
> Also included within the area of infrastructure planning are premises. The huge problems caused by out-of-date buildings unable to support the new technology are well known to most institutions. In planning premises, substantial provisions should be made for the unexpected. The problems of systems integration will be increasingly hindered by physical constraints. The amount of cabling required to support both office automation and the new dealer support networks is immense. Equally, the development of interfaces between the different systems on different hardware will require yet further boxes (protocol converters, for the initiated). A similar approach should be taken to the ordering of hardware; it is an almost universal rule of hardware planning that there is no such thing as a pleasant surprise. In setting the infrastructure, care should be taken at the outset to introduce (and enforce) common development standards and methodologies wherever possible. Finally, the use of fourth-generation languages and other development tools should be planned for quicker development of new systems in future.
>
> A second pragmatic step in planning for the longer term is to initiate projects to bring the existing data within an organisation under better control. It is little short of incredible that most large organisations have only a vague idea of their business relationships across the organisation. At a minimum, the centralisation and verification of customer information through a manual exercise, perhaps through a 'library' function, should be undertaken; once successful, organisational arrangements should be introduced to keep the data current. This data will be valuable when seeking to integrate systems later.
>
> Thirdly, organisational planning for the information technology department should be undertaken, with the ultimate objective of dispersing systems personnel to the sharp end of the business where they gain knowledge of the markets, become personally committed to providing systems for the money makers and are best placed to respond promptly to requests for systems changes.
>
> A fourth action that can be taken now is to design a corporate database which will contain in due course the accounting, customer and external data of an organisation, with links into the underlying systems where transaction information is held. This database will sit at the heart of the later generation of systems and will be the key to integrating these systems to provide the information needed for strategic management of the business. Integration will involve not only building the central database but, equally difficult, designing and implementing interface systems to extract automatically data from the underlying systems and to download the information to the appropriate users through the office network (hopefully) established already.
>
> What comfort is there for organisations confronted by huge investments in information systems and premises which current levels of profitability generally do not support? The only crumb that can be offered is that few organisations, if any, are much better off. There is a strong tendency within the banking industry to believe that others are far advanced of one's own institution in systems. This is not so. Success will lie in the achievement of clearly defined goals.

CHAPTER THIRTEEN

The outlook for international banking in London

The Parkinson–Goodison accord in the summer of 1983 ensured that London will take its rightful place as one of the world's three major financial centres. The City has had to make some concessions, but the real question is whether it has made too many. It is very likely that soon the City of London will no longer belong to the Londoners. To a large extent the writing was already on the wall in the shape of the Euromarkets. The fact that London will now be a true international financial centre should be of some consolation.

The worst possible start to the new markets would be for London to be entering a sustained bear phase in the immediate aftermath of Big Bang, although this would not signal the end of the markets. It would separate the weak from the strong sooner than would otherwise have been the case. A possibly more negative development would be short-sighted punitive action taken against the City of London under a new government, especially if that were a Labour goverment. There is no love lost between the Labour Party and the City, although, ironically, the Stock Exchange has generally performed well under a Labour government. However, Labour threatens to reimpose foreign exchange controls, devalue sterling and heavily tax anyone earning over £27,000 a year, a level considered to be 'substantially' wealthy. This may be political rhetoric, rather than a rational strategy for the future prosperity of the United Kingdom. Denis Healey as Labour Chancellor of the Exchequer promised to squeeze the rich 'until the pips squeak', but the only result was that there are now more millionaires in the United Kingdom than ever before.

The reimposition of withholding tax or foreign exchange controls would set back the City's new markets by years. The recent Australian example confirms that the suggestion of a return to capital markets hindered by restrictions is widely condemned. Any major devaluation of sterling would destroy foreign investor confidence and temporarily torpedo the new gilt-edged market, which may have to rely for its survival on increased overseas purchases.

An excessive rate of income tax on earnings over £27,000 per annum would discourage the incentive to save and wider share ownership, in which the United Kingdom already lags substantially behind its more affluent sovereign credit counterparts. Instead, the United Kingdom should follow the United States and attempt to lower taxes. British Telecom and the Trustee Savings Bank have proved that demand does exist from sources which have, as yet, been largely untapped. It would be foolish to stifle wider share ownership.

The City's financial community has a welcome ally in the Bank of England, which has lobbied both Conservative and Labour governments in the past, usually with success. The Bank may indeed have reservations about control of the City passing quickly from domestic into foreign hands, and the concern can be appreciated. But change was inevitable and the Bank was among the first to sense this. If the City was going to survive, it could not do so in splendid isolation. It had to learn to live with foreign competitors and, if necessary, learn from them. The City also had much to teach others. It was still the insurance centre of the world and it possessed the third largest capital market and the third largest equity market. UK merchant banks were renowned for their innovative skills. London was arguably the leader in international fund management and ranked beside New York in foreign exchange dealing.

The role of the Bank of England will be to encourage these natural strengths and promote new developments. Because of its natural time advantage, there is no reason why London should not be one of the centres of each major financial market. London will never become the centre of trading in US Treasuries, any more than New York will dominate the UK gilt-edged market, but there is no reason why London should not vie to become the second largest US Treasury trading centre after New York. The huge influx of foreign banks into London has slowed dramatically. Two US banks have even closed their London operations. If London can prove beyond any doubt that its position as a financial centre is strong they, and many others, will have to return. A highly successful US swaps team operating out of the West Coast, for instance, could not hope to be totally effective without a counterparty group in London. If time zones are really to become as important as we are being led to believe, London must not only continue its role of investing in international securities, but must also trade them. It represents an opportunity for which London seems singularly well equipped and which should not be missed.

Equities and Euro-equities

The area where the Stock Exchange felt most threatened and fought hardest to protect the interests of its members was equities

Some of the overseas tourist guides confused the Royal Exchange with the Stock Exchange

and Euro-equities. Stockbroking was an enjoyable and usually a very comfortable profession. Some years were better than others, but the good years tended to be very good, and it is more than a decade ago, in 1974/75, that the London Stock Exchange member firms last saw profitability almost disappear.

Much of this consistent prosperity was due to fixed commission rates. When this has changed, so too will the whole character of the Stock Exchange. Those who believe that a sizeable amount of business will still be transacted on the floor of the Exchange are more likely to be traditionalists than visionaries of the new financial markets of the 1990s. When commission rates fall by 30 to 40 per cent, as seems likely, volume is not going to rise by a proportionate amount to compensate for the fall in commission income. Net income will therefore decline and some smaller, less well-capitalised firms will go out of business altogether or will be forced to merge.

The first casualties are likely to be the smaller London-based houses with high fixed overheads, no net market making skills and no research capability to attract institutional orders. A loyal smaller private client base was a fine asset in the past, but investors with portfolios of, say, £25,000–£100,000 are precisely the people who tend to follow share prices quite closely, and they may be among the first to ask where the benefits of reduced commissions are being reflected in their share dealings.

There will be any number of alternative providers of advice who also have execution capability. The high-street banks are launching a major campaign to attract the small investor. Credit facilities, especially for 'stagging' popular new issues such as British Telecom, may be made available to woo new investors into the field of securities. The building societies are countering the invasion of their own territory by banks offering mortgages. Building societies may offer loans for financial services, although it must be questionable whether many are qualified in all aspects of the financial services industry. The 'money shop' concept, such as that offered by Quilter Goodison in Debenhams' stores, attracted considerable attention, but the market for this type of service, despite Quilter's enthusiasm, is unlikely to be significant. In the United States, the experience of installing Dean Witter shops in Sears Roebuck stores has not produced substantial returns for the parent company. Several thousand new accounts have been opened, but most of the new customers have been more interested in buying insurance policies, which Sears Roebuck can also provide, than in shares.

The gilts market of the future must be internationalised and its securities distributed more widely

The real battle in the equity sector will be for the brokers to maintain their share of the institutional business. Sixty or more will be competing for that business. The attitude of the institutions themselves will be crucial. Will they be prepared to pay even a reduced commission when 60 houses are making net prices? What value will they place on research? Will trading volume increase to the levels that some optimists expect? Ron Artus' comment, quoted earlier in the book, suggests that there is no correlation between increased trading and improved performance. Net prices and lower stamp duty will be of little assistance if the stock market is in a serious decline. It will take time before the attendant problems of dual capacity become fully recognised. Culture clashes aside, the practical difficulties of having trader and analyst sitting virtually side by side is bound to cause some concern among the institutions. They feel that the information which they receive cannot be fully guaranteed to be impartial. It is all very well to say that the system has worked quite adequately in New York, but the Americans have more than a 10-year head-start over London, and an SEC which dwarfs its UK regulatory counterparts.

Trading in the 100–150 alpha stocks is likely to be so competitive in the initial stages, that it will be difficult to earn significant profits. The temptation will therefore be to position for the house account, which is dangerous for those houses, including many British ones, who have no previous experience in making markets. The in-house analyst may have a particular view of a share, but the overall reputation and record of analysts is often mediocre; unless clients are not only informed first but also agree with the analyst's conclusion, the share price is unlikely to move to any extent.

It is quite possible that satisfactory levels of profit will be maintained in the beta and gamma shares. There will be fewer net market makers, and where these do exist, the dealing spread between bid and offer prices is likely to be significant. Clients will be prepared to pay for niche research, whereas they feel less inclined to examine thoroughly the merits of, say, eight individual reports on British Petroleum.

Net market makers in the beta and gamma stocks may also find themselves in the dangerous position of being the only market maker, although commentators are probably correct in their opinion that this situation is really no different from the old jobbing system, where it was sometimes impossible to deal in second-rank securities. A key consideration in the City's new capital markets will be the willingness of the major market makers to deal with each other. There will be some dissension; but the Eurobond experience is an encouraging example: the majority of the largest houses regularly trade in huge amounts directly with each other, without having permanent recourse to the brokers.

As the essential role of the London Stock Exchange is to raise new capital, the attitude of the corporate treasurers will also be vital. Will they retain their old loyalties not only to the traditional merchant bankers but also to their traditional methods of raising money? The process of smaller companies coming to the market for the first time will not change and the success of the USM has done much to despel the criticism that the City neglected private enterprise. The rights issue mechanism will not disappear, but it now faces very serious competition from Euro-equities and equity-linked Eurobond issues, which are little more than thinly disguised rights issues. The foreign houses in London already control most of the Eurobond market. They intend to gain the same position in Euro-equities and have already left most of their British counterparties at the starting-gate. The foreign houses' enthusiasm for Euro-equities is not purely for publicity or league table position. A major consideration is profit. Profitability of Euro-equity issues is substantial, and is needed to replace the declining earnings from their primary straight Eurobond operations.

For the larger UK companies the rights issue mechanism must be seen as increasingly cumbersome, and perhaps also expensive. Euro-equity offerings should result in wider international ownership. It will also assist UK management to understand better the attitude and requirements of overseas investors. At the present time foreign ownership of UK equities, with the exception of the US institutions, is far below the levels which they merit. If UK management explained their companies better to foreign institutions, a prerequisite for any Euro-equity offering, the proportion of foreign ownership might rise. The question of Euro-equities and pre-emptive rights of existing UK shareholders still needs to be fully resolved, but some progress has already been made. The Euro-equity is not a temporary development and it is important for the Stock Exchange and the UK merchant banks to claw back the lead which has already been established by foreign institutions.

The gilt-edged market

There is almost universal agreement that profitability will be very limited for many of the primary gilt dealers in the early stages, except in the event of a major bull market, and meagre thereafter. It is no secret that more than two houses would have liked to have followed Union Discount's decision to withdraw as a primary gilt dealer, but it does appear that some 'behind the scenes' pressure was exerted by the Bank of England. Profits for the remainder will be made by correct positioning. Because positioning involves direct risks, some major losses will inevitably be incurred. Most of the houses have the financial backing and the patience to see a bad period through, but there will be those who will rue their decision to join the gilt 'club'.

Within the next four years, however, the number of primary gilt dealers will have probably increased rather than contracted. All the four major Japanese securities houses will become primary dealers. The gilt-edged market is large, efficient and highly liquid – key considerations for the Japanese houses and their enormously wealthy domestic clients. While they would obviously wish to see sterling as an appreciating rather than a declining currency, this may not be as important a factor as some commentators have suggested. Their purchases of billions of dollars of US Treasury bonds, for instance, have coincided with a period of unprecedented weakness of the dollar.

The Bank of England's concern that the gilt-edged market, which funds the British government's public sector borrowing requirement, may move largely into foreign-owned hands deserves sympathy. However, the gilts market of the future must be internationalised and its securities distributed more widely than to a domestic audience. The Bank of England will, no doubt, continue its exercise of prudent control and influence, but the 'raised eye-brow' of yesteryear may go unnoticed by the partner of a major Wall Street investment bank or a US commercial bank. Nevertheless, the commitment of the primary market gilt dealers to a market in which there may be no early profits, and little to come later, shows commendable resolve and augurs well for the world's third largest capital market.

The Eurobond market

Any major capital market which does not model itself on the 'Euro' model will find itself quickly outdated. London stockbrokers did resent the success as well as the salaries paid to the Euro 'upstarts'. Instead they should have been grateful. 'Euro' became synonomous with progress. The City of London required progress and the presence of the Euromarkets provided the perfect model for imitation without the time-absorbing necessity for innovation. It was already a blue-print for success.

It is already too late for most Stock Exchange member firms to become important participants in the Eurobond market. They all had their chance more than 20 years ago. The cost of entry is not prohibitive, but it is doubtful whether it would be productive. The fact that the Euromarkets are, mistakenly, considered to be unregulated does cause concern to those whose sole professional career is regulation. The unfortunate result of professional regulation is bureaucracy. If bureaucracy is allowed to permeate the hugely successful Euromarkets, it will inhibit growth and force sectors of the market to seek a new permanent residence other than London. No effort should be made to hinder the progress of a market which raises more than US$200 billion annually with extreme efficiency. The regulators would be better to spend their time looking at the ancilliary financial aspects which the growth of the Eurobond market has spawned. Six or seven years ago, swaps had not been invented. Today 80 per cent of all conventional straight Eurobond issues are swap-driven. While there has been much discussion, few conclusions have been drawn on how these off-balance sheet financings should be restricted or how they should be accounted in assessing a bank's credit quality. London can thank the Eurobond market for making it a power in the now huge swaps industry.

The question of off-balance sheet financing is delicate and should be left to the central banks to whom the overseas subsidiaries are ultimately responsible. The day-to-day regulation of the Euromarkets can be left in the perfectly capable hands of the International Securities Regulatory Organisation (ISRO), the Association of International Bond Dealers (AIBD) and the International Primary Market Association (IPMA). As the Eurobond market has been at pains to point out, it is a professional market dealing with professionals. Authorities, such as the Securities and Investments Board, should confine their early attention to those industries where the clientele is non-professional.

If the Eurobond market has a problem, it is of its own making. Perhaps inadvertently it has become a borrowers', as opposed to a lenders', market. A similar situation may take place in London's equity markets when fixed commissions are abandoned and the

institutions find themselves in almost total control. It is not at all easy to find accurate details of Eurobond profitability but it would almost certainly be fair to say that primary market lead management profitability has declined very sharply. The old adage used to be that one good lead management made up for six poor co-managements, but today most straight debt issues lose money before allowing for swap fees, and the professional skills of the co-managers are assessed not by the numbers which they accept, but by those which they turn down. In recent years the profits of Euro houses have come from equity-related issues, floating rate notes, secondary market trading, new product innovation, swaps, futures and outright positioning. The pendulum has swung much too far in favour of the borrower, who can start by asking for terms which seem absurd and finish with ones which are even more advantageous. If all the borrowers concerned were triple-A-rated sovereign credits, it would justify the pre-occupation with league tables which are the natural preserve of no more than half-a-dozen houses. The problem arises when a medium-quality single-A-rated borrower with a regular demand for funds can ask for, and in most cases receives, competitive bids from some of the world's most powerful institutions.

So much for the traditional loyalty which many London Stock Exchange member firms expect from their clients in a deregulated market. If the Eurobond experience has taught practitioners anything at all, it is that price counts before anything else. However, unless borrowers accept more realistic terms, they will watch the market contract. Euro houses are seeing their profits come under pressure for the first time in several years. The Euromarkets houses introduced sky-high salaries to the City, and the cost of communications and new technology has resulted in very high fixed overheads. In such circumstances it is doubly absurd to offer new issue terms to clients who will offer no enduring loyalty and who may easily cost the house a considerable amount of money for the sake of winning a mandate.

The great strength of the Euromarket is its flexibility and its record of innovation. Some innovations have been considerably better than others, but many created a welcome, although often brief, flow of profits. The market's consistent advantage, matched by no other capital sector, is that it is all things to all men at any time. When US dollar bonds are out of fashion, the Euromarket can offer ECUs, DM debt warrants, Swiss franc zero coupon bonds or any other investment which is popular at the time. That is why

The Stock Exchange seen from Old Broad Street

The UK houses with proven skills in specific overseas equity markets should extend that expertise

those who predicted the market's early demise are now beginning to appear very foolish. London's only regret should be that 20 years ago this market was theirs for the taking.

Domestic corporate and non-sterling markets

As has been said, London's capital markets have the advantage of unlimited flexibility, and practitioners should be quick to capitalise on their good fortune. It is unlikely that the UK corporate bond market, dormant for much of the past decade, will appeal to other than domestic investors. The attractions of Eurosterling and Bulldog bonds have been outlined in detail in Chapter 4. Foreign investors will buy these issues, but these investors' focal point of interest is likely to be in trading in the gilt-edged market.

The easiest mistake which the City could make in the new capital markets will be to concentrate expansion strategy and in-house expertise solely on domestic sterling securities. The London houses must learn to trade and offer advice in Deutschmarks, French francs, guilders, ECUs, Canadian dollars, yen and dual-currency bonds. The trading apparatus is already in place. Familiarity with the non-sterling instruments is all that is now required. If New York's Salomon Brothers can rank within the top six dealers in Japanese yen-denominated government bonds in Tokyo, why should not Barclays de Zoete Wedd or County Bank rise to be the second or third most important market maker in ECU bonds? The identification of potentially profitable niches will prove to be far more rewarding than, say, trading Eurosterling securities along with most of the 27 primary gilt dealers. Robert Fleming has already proved that it is possible to compete head-on with the Japanese houses not only in Tokyo but also in London. The UK houses with proven skills in specific overseas equity markets should extend that expertise first to equity-related bonds and then to straight debt securities. The ambitions should be similar to those of Shearson Lehman, who plans to be directly represented on every major European stock exchange within a few years. When this has been achieved, Shearson Lehman will be able to offer its customers true international portfolio advice and recommend geographical allocation of assets.

The smaller London and regional firms will have to learn the new rules, which are to internationalise or face the prospect of much harder times ahead. It is not difficult to persuade the average British private investor, whose assets and expenditures are all in sterling, that he or she should buy UK equities or gilts. Probably the investor is reasonably pleased with the results. What was almost certainly not explained to the same investor is that if he had placed all his surplus cash in Matsushita Electrics or Sony, he would have achieved minimum probable returns of 50 per cent per annum including currency appreciation. Despite the undisputed professionalism of many London stockbrokers, they suffer from blinkered currency vision in favour of sterling. As sterling cannot be regarded as a historically strong currency, a proportion of assets should be held in non-sterling securities or in UK companies with substantial overseas income.

The UK merchant banks will have to move quickly to extend their dealing and distribution capabilities in non-sterling bond issues. At the moment they seem to be falling further and further behind. It should be very disturbing to see UK mandates being awarded to foreign houses. A classic recent example was British Telecom, whose enormously successful share flotation had been lead managed by Kleinwort Benson. However, when British Telecom decided to tap the Euromarkets, the lead manager was not Kleinwort Benson but Morgan Guaranty Ltd. When one of Britain's best known companies chooses an American bank to manage a new financing, it should serve as a warning for the whole of London's business community. The new financial markets were urgently needed to replace a system that was hopelessly antiquated. History will probably record that the three-and-a-quarter-year time gap between the Parkinson–Goodison agreement in July 1983 and the ending of fixed commissions in October 1986 was insufficient for the City to adjust. It is only hoped that history will also record that the changes should have taken place years before and if, as seems likely, a substantial portion of the new financial markets fall under foreign control, the City and the Stock Exchange have only themselves to blame.

THE SPONSORS

Alexanders Laing & Cruickshank – a new international securities house

Mercantile House is an international financial services group consisting of three operating divisions: Wholesale Broking, Investment Banking and Securities Trading and Fund Management. These divisions between them cover the world's financial markets from retail to wholesale, but it is the international banking and securities division that is particularly involved with the current restructuring of the London equity and government securities markets. This division operates under the name of Alexanders Laing & Cruickshank and will cover all the securities markets in the principal financial centres of the world.

The members of the group

Alexanders Laing & Cruickshank can trace its history back to 1810 when William Alexander began his bill broking business near The Artichoke, a coffee house in Lombard Street that was a popular meeting place for financiers in the early nineteenth century. The Artichoke is remembered today as the trademark of Alexanders Laing & Cruickshank. In addition to Alexanders Discount the other constituent companies are Laing & Cruickshank, members of the Stock Exchange, formed in 1882, R. J. Rouse, commodity brokers and dealers, founded in 1847, and Saturn Leasing, which began business in the 1970s. Together these companies form the group of Alexanders Laing & Cruickshank Holdings; they were put together by Mercantile House between 1984 and 1986 to form the basis of a new international securities house. The necessity to create such a securities house stemmed from two factors: first, the agreement between the government and the Stock Exchange in July 1983 that triggered off the chain of events which has led to the current fundamental changes to the structure of the London markets; and second, the fact that the previous system of specialist single capacity and individual ownership of Stock Exchange firms together meant that no broadly based securities houses existed in London prior to 1983. Mercantile House, like all its competitors, was thus forced to create a securities house by assembling the component parts.

These component parts brought together the skills of trading and market making, distribution and origination, and knowledge and experience of both the equity and government securities markets. It was the requirement for these skills that led Mercantile House first to acquire two discount houses (Alexanders, and Jessel Toynbee and Gillett) and then Laing & Cruickshank, and to join them with two of Mercantile House's existing subsidiaries, R. J. Rouse and Saturn Leasing.

Expansion of services

In 1986, following the acquisition of 100 per cent of Laing & Cruickshank once this was permitted by the Stock Exchange, Alexanders Laing & Cruickshank Holdings was finally put together and organised in its present form. Alexanders Laing & Cruickshank Holdings (ALCH) today consists of four international product areas: fixed interest securities and futures, equities, corporate finance, and investment management. In order to extend the coverage of the international markets ALCH has in 1986 opened a representative office in Tokyo and acquired May & Mellor in Australia. These companies have been integrated into ALCH which now operates through an international network of offices in London, New York, Tokyo, Hong Kong, Singapore, Sydney, Melbourne and Kuwait.

In the UK the retail stockbroking network has been strengthened by the aquisition of Wishart Brodie & Co in Edinburgh, Carr Workman, Patterson, Topping & Co in Belfast, and the opening of an office in Worthing. To this expansion of services can be added the formation of Heritage Personal Finance, which offers financial services particularly to estate agents, and the group's participation in Household Mortgage Corporation, which specialises in the origination and securitisation of domestic mortgages.

Through its four product areas and international network ALCH has thus been carefully created to be one of the new generation of securities houses taking advantage of the opportunities offered by the accelerating pace of change that is currently being stimulated by deregulation, advancing technology, the growth of the futures and options markets, and the securitisation of assets.

The product areas

Fixed interest securities and futures, managed by Jeremy Hardie, is the first product area. It is itself divided into three corporate entities. These separate entities are currently required by the authorities but it is expected that over time this requirement will be

Above: Jeremie Hardie manages fixed interest securities and futures. Right: Gerry Mordaunt, responsible for equities

Left: Mark Evans, head of corporate finance. Above: Mark Powell, responsible for investment management

relaxed and the companies will merge. Together these companies cover every aspect of the international fixed interest markets.

Alexanders Discount carries out its traditional role, taking deposits, dealing in government securities and acting as a market maker in treasury bills, CDs, bankers acceptances, and other short-term money market instruments.

Its activities, mainly at the short end of the market, are complemented by Alexanders Laing & Cruickshank Gilts Ltd, which will be one of the new primary dealers in government securities when the new market structure comes into effect. It has integrated the gilt sales team from Laing & Cruickshank with a team of traders and thus combines the two essential characteristics of market making and distribution. The activities of Alexanders Discount and Alexanders Laing & Cruickshank Gilts Ltd operate from a single dealing room and they are coordinated with the business of Alexanders Rouse, who act as broker/dealer in the world's futures and options markets.

The services of Alexanders Rouse cover currencies, financial instruments, stock indices, metals and agricultural products, and a specialised options unit handles both cash and futures options. Alexanders Rouse has been a pioneer of 24-hour trading in the world's financial markets and its London and New York offices are staffed from the opening of the Australian markets until the close of the west-coast US markets. This 24-hour worldwide coverage will be extended by ALCH to the world's fixed interest securities markets and, in particular, the Tokyo office and the recently formed Alexanders May & Mellor in Sydney and Melbourne, complete the international circle.

The equity area is the responsibility of Gerry Mordaunt and functions under the name of Alexanders Laing & Cruickshank Institutional Equities Ltd. This operates through its London office and also works closely with May & Mellor Laing & Cruickshank in Australia and the Tokyo office. The institutional department has a client base consisting of the majority of significant UK investment institutions as well as a large number of institutions in other parts of the world. It is supported by a strong research team and works closely with the corporate finance department. It is also particularly active in investment trusts and the international mining markets where Laing & Cruickshank has maintained close connections for many years in North America, Australia and South Africa. It has established a strong presence in the new traded options market which is a fast growing sector of the London market. Once the new system permits, these existing skills will be joined by a recently recruited team of market makers and dealers, and together these will enable Alexanders Laing & Cruickshank Institutional Equities to offer a complete service in the competitive new market environment, enabling it to function as the complete broker/dealer and block trader.

The corporate finance area is managed by Mark Evans and covers a number of skills. The first is general corporate finance, including the sponsorship of company flotations on the Stock Exchange, raising new capital through issues of shares and other securities, executing take-over offers and advising offeree companies on defence strategies, and providing corporate clients with general advice on financial matters. The mergers and acquisitions department concentrates on advising clients on acquisitions, management buy-outs, and the provision of venture capital. Alexanders Laing & Cruickshank Asset Finance Ltd (incorporating Saturn Leasing) provides consultancy and broking services to the UK public and private sectors on leasing and other ways of financing assets.

The investment management area comes under Mark Powell and is the retail part of the group, operating under the name of Laing & Cruickshank Investment Management Services Ltd (LACIMS). Laing & Cruickshank has traditionally been strong in its services to private investors and regards portfolio management for individuals, charities, trust funds and pension funds as a very important specialisation. The firm is now probably the largest private client stockbroker in the United Kingdom with offices in London, Belfast, Eastbourne, Edinburgh, Glasgow, Jersey, Taunton and Worthing. It provides private investors with investment advice and information on an individual, personal and flexible basis. The firm is able to draw on its extensive research facilities for information and advice and on its internal computer systems for instant valuations and portfolio records information. Comprehensive advice on inheritance tax, pensions and other non-Stock Exchange investments are provided by a subsidiary company, Laing & Cruickshank Financial Services Ltd. To this can be added the provision by Heritage Personal Finance of mortgage, insurance and other financial services. From Reading, Heritage intends to complement Laing & Cruickshank by providing the retail customer with an additional range of services through a network being established in conjunction with local firms of estate agents who feel unable to provide a full range of services from within their own resources. This extends Laing & Cruickshank's exposure to the retail market, and the exposure is carried further by Household Mortgage Corporation which offers a competitive mortgage facility for the individual householder as well as providing an asset-based securities product that will be traded by the wholesale market.

The future

In the 176 years since the formation of the first of Alexanders Laing & Cruickshank Holdings' component companies there have been some dramatic changes to the world's political, social and economic structure. In the last 30 years the financial markets have undergone the most far-reaching period of change ever seen, due principally to the volatility of interest and exchange rates, advances in technology and the advent of the Eurodollar market, which together have turned the world into a single financial market place. The next 30 years are likely to see even greater changes and Alexanders Laing & Cruickshank Holdings will continue to develop its four product areas to respond to the challenges presented by the evolving securities markets of the world.

No.8 Bishopsgate, head office of Barings

Playing Big Bang for the long term – Barings plc

Barings, founded as Baring Bros. & Co. in 1762, is the oldest merchant bank in the City of London. Since its foundation its equity has been privately owned and it has remained wholly independent. Its original business was essentially international; its strengths and outlook have remained strongly international and it has consistently maintained a high level of business with sovereign nations and their central banks and with multinational companies. By current standards it appears to have had almost an aversion to publicity.

Baring Brothers & Co., Ltd. (**Baring Brothers**) was formed in 1891 to carry on the business of the bank which had up to that date been owned by a private partnership. In November 1985, Barings plc (**Barings**) acquired the whole of the share capital of Baring Brothers and became the parent company of the group. The operating companies of the group are now centred on Baring Brothers, which coordinates the merchant banking, capital markets and securities trading operations, and **Baring Investment Management Holdings Ltd.**

In recent years, in addition to a general expansion of its traditional merchant banking business, the Barings group has significantly increased the range of its activities. Recently established operations or acquisitions include:

- **Baring Securities Ltd.**, a broking and dealing business originally established within stockbrokers Henderson Crosthwaite, specialising mainly in Japanese securities;
- Wilson & Watford (now **Baring Wilson & Watford**), a London jobber specialising in gilts, Bulldogs and debentures;
- **Baring Brothers S.A.** in Geneva, a subsidiary with activities in capital markets and investment management;
- **Baring Fund Managers Ltd.**, a unit trust management company.

Barings and Big Bang

Big Bang has effectively broken up the traditional structure of the City of London and no merchant bank has been able to ignore the vital changes in function it has brought about. The major merchant banks have not, however, adopted a uniform strategy towards Big Bang and their varying approaches will, before long, produce institutions which differ markedly from one another. Size alone will decline in importance as a distinguishing feature.

For Barings, as for other major merchant banks, the debate is all about how to take the greatest possible advantage of the opportunities presented by Big Bang, given the very strong position from which Barings is able to approach any new City regime. Against this background, Barings has adopted certain key points of policy which are reflected in the steps which have been taken so far:

- In the first place Barings will continue to benefit substantially from remaining independent and privately owned.
- The advent of Big Bang is seen by Barings as an opportunity to extend its existing activities to a substantial extent. It has not been tempted to make any very major acquisitions; it has, however, made a series of strategic acquisitions. It is also expanding laterally, particularly in its broking, dealing and market making operations.
- Barings' expansion in dealing and market making recognises that the shape of the securities industry after Big Bang will require any active London issuing house to have far greater capacity in these areas than would have been necessary in the past.
- Growth and development on a secure and long-term basis is impossible unless Barings can offer a working environment which will attract and retain high-calibre staff at every level. Everything that goes to make up the working environment – including in particular management style, remuneration, internal competition and communications – must be a constantly open topic and must be sufficiently flexible to accommodate a number of different working styles. A standardised approach to management will often only inhibit the integration it is supposed to achieve.
- Financial reporting and control is far more demanding and is of higher priority than in the past, partly because of the new kinds of asset which any competitive merchant bank must now be prepared to deal, partly because of rapid growth and expansion and partly because operational costs are changing in form and rising sharply.
- There are some significant new risks associated with Big Bang. It is, for example, likely that some firms operating in the securities industry will not succeed, particularly when the current phase of strong securities markets worldwide comes to an end. The appearance of new and inexperienced players competing for positions in the securities markets will make the markets themselves less stable and less efficient.

■ The changes stimulated by Big Bang are unlikely to have run their course by the time the new structure of the securities industry is in place. There may well be acquisition and development opportunities in the future which will be more attractive than those which have been available in the run-up to Big Bang.

Historical

The finance of international trade – first in wool but rapidly extended to other commodities – led Barings naturally to the provision of banking facilities for overseas customers. It also created contacts with foreign governments and established relationships which led to the firm's involvement in raising funds in London for the development of overseas countries.

In the early nineteenth century, this activity centred on the United States of America, Canada and Argentina; in particular it included financing the purchase of the Territory of Louisiana by the United States from France in 1803. In the same year, Barings became European financial agent for the United States Government, and in 1818 for the Bank of the United States. The first Canadian loan was made in 1835, and over the years funds raised by Barings in London helped to finance the development of North American industry and public utilities.

Trade and banking links with South America were strengthened when the Latin American countries achieved independence in the early nineteenth century and Barings issued the first Argentine loan in 1824. Barings was closely involved during the remarkable economic expansion in Argentina and Uruguay in the 1870s and 1880s and in seven years issued loans totalling almost £40 million for the two countries.

Around the middle of the nineteenth century the firm's business in Europe, the East and Australia began to assume substantial proportions. In 1817, Barings had arranged the loan to France for reparations after Waterloo, and contact with Russia begun in the 1820s led to the firm's appointment in 1846 as London financial agents for the Imperial Russian Government. This appointment lasted until the fall of the government in 1917 but Barings continued to be custodian of the Russian balances in London until the UK–Soviet agreement was signed in July 1986.

The firm had been involved in trade with China since 1833 and when the Japanese ports re-opened to British merchants in the middle of the century, Japan afforded a new market to explore and develop. Barings later underwrote bond issues in London for both the Chinese and the Japanese Imperial Governments.

Barings' business in 1986

In its role as a recognised bank and a member of the Accepting Houses Committee, Baring Brothers is active in all the traditional areas of banking both in London and, through **Baring Brothers Asia Ltd.**, in Hong Kong. In recent years this business has been broadened to include interest rate and currency swaps and options, forward rate agreements, sterling bill issuance facilities and other new products. In June this year Baring Brothers initiated a $500 million multiple-option facility for Allied-Lyons plc which enabled them to access financing at their option by way of Eurodollar notes, sterling commercial paper or multi-currency cash advances.

Baring Brothers has achieved a leading position in the Eurosterling and Bulldog markets and already has a significant dealing presence in these and other markets. Baring Brothers has made a substantial contribution to the development of the sterling public debt markets since their reopening in the early 1980s and has been responsible for a number of innovations, most recently in bonds associated with property development. In 1985 Baring Brothers lead managed 11 sterling issues, totalling £800 million, including three Bulldog issues, and was also involved as co-manager in a further 38 sterling issues, having a total principal amount of over £3 billion. A similar level of activity has been maintained through the first half of 1986.

Baring Brothers is an active manager of Eurobond debt issues in other major currencies in both the fixed and floating rate sectors. Management of international equity-linked issues, notably Japanese and European convertibles, represents an important and expanding part of the group's primary issue business.

Over the last three years Baring Brothers has considerably strengthened its capacity to trade and distribute debt securities – principally Eurobonds and Bulldogs – in volumes appropriate to its primary issuing business and its anticipated operations in the post-Big Bang era. In the run up to Big Bang the Treasury and Trading division has been the area of most rapid internal expansion within the group; its operations are effectively extended by those of Baring Wilson & Watford in the gilt and debenture markets and by Baring Securities Ltd., which now trades out of London, Hong Kong, Tokyo and New York, primarily but not exclusively in Japanese and other Far Eastern securities and warrants.

In corporate finance Baring Brothers has shared in the increased activity of recent years and is now busier than it has ever been. In addition to a significant share of conventional corporate finance business and new issues, Baring Brothers has always been strong in ultra-confidential advisory work. Recent assignments have included advisory work for UK government departments and institutions.

Baring Brothers has been involved in a series of important transactions associated with the approach of Big Bang. Baring Brothers has acted among others for Vickers da Costa, Scrimgeour Kemp Gee, Union Bank of Switzerland (in its acquisition of Phillips & Drew) and Royal Bank of Scotland (in its acquisition of Charterhouse Japhet).

The corporate finance services offered by Baring Brothers are enhanced by operations in Tokyo, New York, Geneva, Hong Kong

The Directors of Barings plc: standing (left to right), Robin Broadley, Miles Rivett-Carnac, Robin Dent and Andrew Tuckey; seated (left to right), Nicholas Baring, Sir John Baring (Chairman) and Peter Baring

(where it recently led the Cathay Pacific flotation), Singapore, Kuala Lumpur, Seoul and Sydney, where **Baring Brothers, Halkerston & Partners Ltd**. has established itself as the leading Australian corporate finance house.

Baring Investment Management Holdings Ltd. through its subsidiaries manages funds totalling (at the end of June 1986) some £8.7 billion for pension funds, charities and endowments, governments, central banks, captive insurance companies, offshore and authorised unit trusts, investment trusts and private individuals. The group has developed particular expertise in the markets of the Pacific Basin and North America. As well as a large London-based investment organisation, the group has fund management companies or offices in Boston, New York, Hong Kong, Tokyo, Geneva and the Channel Islands.

Baring Fund Managers Ltd. was set up in 1985 to develop the previously limited and specialised involvement of the group in UK unit trusts. Since its formation, Baring Fund Managers Ltd. has promoted a total of 12 funds with an emphasis on Japanese and other Far Eastern securities.

Through its interests in **Landauer Associates Inc**. a leading American real estate consulting firm, and **Baring, Houston & Saunders**, UK property consultants, the group has expanded its capacity to advise clients in relation to the acquisition, financing and realisation of major properties and other real estate matters.

Outlook

The Barings view of Big Bang was summarised recently by Sir John Baring, Chairman of Barings plc:

'The shape of our business and the number and emphasis of its activities have been changing steadily over the last decade. Even before the most recent changes which have taken place, someone who retired 10 years ago would find many aspects of it altered beyond recognition.

'We are now, however, part of a further process of change formally initiated by the government/Stock Exchange agreement of 1983 which affects all those who are, like us, active in the securities business in London. It is safe to say that the changes now initiated in this type of business together with those yet to come are on a scale and will take place at a speed which will justly be characterised as revolutionary.

'We have believed for some time that the way in which securities markets have been developing internationally over the last decade, combined with the remorseless decline of margins on banking business, made changes of the sort that are now occurring inevitable sooner rather than later. In the process some of the good characteristics of the old system will be lost, but we see no point in looking backwards and we are confident in taking up our selected areas of activity in the new order.'

Dow Financial Services goes Royal

'More business, less government'. In the eyes of Prime Minister Margaret Thatcher, this is evidently the correct tonic for London if it is to retain its eminence as a financial centre. Her free-market orientated cabinet encouraged the decision that triggered London's Big Bang on 27 October 1986. On this day, so it seems, new life will be breathed into the hallowed halls of the London Stock Exchange. Fixed commissions, and the barriers between jobbers and brokers will be formally swept away. Foreign financial institutions as well as domestic banks will become much more dominant. Each has sought to establish its presence in the Stock Exchange arena as the rules governing membership have been radically relaxed.

The new freedoms to be enjoyed in London must be considered inevitable. Thanks to modern electronic communication systems, London finds itself in a unique geographic position. It is possible to mould national capital markets into one global market, and London appears to be about to enhance its position at the very centre of the world's financial markets.

After Big Bang, financial services that are orientated towards the domestic British market will have to fight to retain this domestic base, just as they will need to fight for a share of the hotly contested world financial market. At first glance, American-led financial 'supermarkets' would seem to be calling the shots. For years they have been adopting every profitable idea and marketing it in the form of a new product or an even more sophisticated financial technique.

Size and power seem to march arm-in-arm in the revolution. With their aggressive marketing, the North Americans have given many customers what counts in the global financial market – a less costly global service but, above all, profit and performance.

It is more than possible that three categories of financial institutions will survive on the world financial market: First, strongly capitalised, globally organised banks and financial supermarkets; secondly, specialised financial consultants and corporate finance experts, the 'boutiques' now well established across the US; finally, the small, well-organised financial services groups, operating on a worldwide basis, marketing services that offer profit and growth, but offering some unique element in their services that will ensure their prosperity and survival.

Origins and development of the group

The Dow Financial Services Group belongs to this third category. It has approximately 800 employees worldwide and total assets of US$2.5 billion. Its origins go back to 1965, when the Dow Chemical Company created Dow Banking in Zurich. In 1970, Dow Banking opened a representative office in London, which received a full banking licence from the Bank of England in the autumn of 1971. For nearly 10 years the development of the London bank remained moderate before it began its real growth: in June 1981, Dow Scandia Banking Corporation was established, with Dow Banking taking an initial 52 per cent interest, in conjunction with Sundsvallsbanken (Sweden), Forretningsbanken (Norway) and the Bank of Helsinki (Finland). In November 1981, the 150-year old publicly quoted Arbuthnot Latham London merchant banking group

Left: the office of Dow Banking in Zurich
Right: the London office

was acquired. The merger caused the bank's assets to rocket to £400 million in sterling terms, from less than £25 million in 1971. Stockbrokers Savory Milln was partially acquired in 1985 and 100 per cent owned by early 1986.

The flagship of Dow Financial Services is Dow Banking of Zurich, 77 per cent of whose capital is held by the group, ensuring itself a firm foothold in the financial capital of Switzerland and one of the focal points in the world's financial scene. The opportunities offered by Switzerland are further complemented by the bank's subsidiary in Grand Cayman. Dow Financial Services also operates in Singapore through ALASIA, the first merchant bank established in the People's Republic of China and in Hong Kong, and in Kuala Lumpur through joint ventures with the MBf group, the largest financial services organisation in Malaysia.

The sustained growth that Europe has experienced over the past years is reflected in the success of both the Zurich and London operations of the group. It is based on rapid developments in technology and a political environment that accepts industrial efficiency and private enterprise and, not without coincidence, has witnessed the dismantling of regulatory and fiscal structures. The re-emergence of the private investors, who are turning out in force and willing to venture into new instruments of an increasingly liquid capital market, is the most significant development of the 1980s. It is against this background that the group has structured its expansion. Its President, Leslie G. Merszei, has been able to build a financial group that now serves its customers in more than 60 countries.

Royal Trust of Canada

In April 1986, Royal Trustco Limited, the largest and most profitable trust company in Canada, announced that it had agreed with the Dow Chemical Company to the acquisition of the Dow Financial Services Group.

'The purchase of the outstanding organisation of the Dow Financial Services Group will help us achieve one of our strategic goals: to offer a comprehensive range of financial services around the world,' declared Michael A. Cornelissen, President and Chief Executive Officer of Royal Trust.

Royal Trust itself, founded around the turn of the century, today manages assets of over Can$60 billion, making it one of the leaders among comparable institutions. For 1985, it realised a profit of Can$113 million on gross earnings of nearly Can$2 billion, 33 per cent more than in the previous year. The profit for the first quarter of 1986 was Can$37 million, Can$10 million more than in the corresponding quarter a year earlier.

The experts agree that sufficient capital is one of the most important prerequisites for survival in the world's financial markets. The capital power of Royal Trust will further strengthen Dow Financial Services Corporation, soon to be operating under a new corporate name reflecting the change in ownership, and put it in an even better position to withstand challenges and to keep up its dynamic growth.

Worldwide financial services

The proven record of the group in its provision of worldwide financial services shows not only how well they are tailored to meet the needs of the customer, but also reflects the skills of a professional team of high quality.

Dow Banking Zurich, under Chief Executive Arthur Bolliger, reported net profits of Sfr15.3 million on total assets of Sfr2.7 billion

for 1985. The bank is successfully engaged in the capital markets and is undergoing tremendous growth, especially in the area of investment counselling. In the money management field, the services for exacting private customers are being continually expanded. The customer base is showing extraordinary growth which it is believed owes as much to aggressive marketing as to customers' own referrals.

Royal Trustco Limited has completed the sale of Arbuthnot Latham Bank Limited to the Swedish-based bank, Nordbanken. Under the terms of the agreement Royal Trustco receives US$27.8 million in cash, the transfer of certain property interests and the outstanding share capital (28 per cent) owned by Nordbanken in Arbuthnot Savory Milln Holdings Limited. Arbuthnot Savory Milln Holdings Limited, following completion, is 100 per cent owned by Royal Trustco, which acquired the initial 72 per cent interest in June 1986 from the Dow Chemical Company. Arbuthnot Latham Bank was 100 per cent owned by Arbuthnot Savory Milln Holdings Limited.

The sale heralds a major reorganisation of Royal Trustco's UK interests. These interests will be centred around three core businesses. These are:

The Royal Trust Company of Canada
A recognised bank in the UK, established in London as a branch office in 1929 and incorporated in England with unlimited liability in 1955, Royal Trust employs 200 people and has its head office in the City in Cannon Street, with other offices in Manchester and Ipswich. At 30 November 1985 it had shareholders' funds of £39 million with balance sheet footings of £832 million, which today stand at over £1 billion. The bank provides a wide range of corporate banking services to the business community, as well as personal financial services to individuals through its wholly-owned subsidiary Royal Trust Financial Services Limited.

Royal Trust International Asset Management Limited
Formerly Arbuthnot Investment Management, it consists of various companies engaged in worldwide investment management, not only in London but also in the Channel Islands and Singapore. Total assets under management in the former Arbuthnot stable are £250 million and the new company, which will also incorporate the investment management business contained within Royal Trust Company of Canada (London) (£250 million) and Royal Trust Jersey (£300 million) will mean that the asset management business will have some £800 million of funds under management.

Savory Milln Limited
In the UK, Savory Milln, under Chairman and Chief Executive Bob Erith, specialises in providing institutional investors with a high-quality service in certain key areas of the market where it has been established as a leader for many years. It has long been rated top in analysis of construction shares, and among the first three in building materials, engineering and financial stocks.

In the past 20 years, Savory Milln has built up its coverage of continental European markets; to date it is considered the leading UK broker in France, and it is also a leader in Germany. Clients are kept informed of investment opportunities on a worldwide basis through the monthly Global Investment Strategy document covering economies, currencies and markets of 18 different countries.

The highly qualified research and sales team is complemented by London's largest European market making team which makes markets in over 250 continental European stocks. Following deregulation, this will be extended to making markets in UK stocks of the principal specialisations and running a 'leader' book in internationally traded shares.

This unique service has resulted in a rapid growth in institutional clients from outside the UK, whose numbers have doubled in the past year, and has led to Savory Milln being voted one of the world's best brokers, and to offices being established in New York and Singapore.

Dedication and commitment

In the rapidly changing world environment for money management, the group believes that efficiency and service are paramount.

Above: Arthur Bolliger, Chief Executive of Dow Banking/Royal Trust Bank Switzerland, Zurich. Right: Robert F. Erith, Chairman of Savory Milln Limited

Against the apparent market norm, the group believes that the client comes first and the whole emphasis of the group is to continue to remain 'client driven'.

Obviously, this requires a far-ranging commitment to the development of sophisticated computer systems, and much has been achieved in the past 12 months. The clients, which include private individuals and corporate and pensions funds – as well as a range of mutual funds both on- and offshore UK – can continue to expect the degree of dedication that the group has gained a reputation for over the past few years. The Investment Division has recently appointed a new Executive Chairman, Adrian Collins, whose previous achievements are well documented, and underline the commitment to establishing a major international asset management business.

'Our emphasis will be in providing a first-class service to these private investors, while also expanding our capabilities to serve our institutional clients better', comments Collins in assessing his division's future strategy.

Through other subsidiaries, the group offers sophisticated financing in the fields of leasing and international trade. The growth of factoring in Europe, and in the UK in particular, is mirrored in the success of the group's own factoring and invoice discounting operations, which have grown dramatically over the last two to three years, thus broadening the group's overall marketing and communications base.

ALASIA was originally established in 1970 as a joint venture between Arbuthnot Latham and the Standard Chartered Bank. Arbuthnot Latham Bank owns 60 per cent and Dow Financial Services Corporation 40 per cent.

Apart from its well-established merchant banking activities ALASIA, led by Managing Director Laura Hwang, also manages the Singapore Growth Fund and the Pacific Growth Fund, allowing participation in the stock markets of Singapore and Malaysia. It is also active in corporate financing and the regional merger and acquisitions areas.

In Hong Kong, Dow MBf Limited, under the guidance of its Managing Director, Mark C. Blacker, operates as one of the limited number of deposit taking companies (LDTC). The company's recent major successes were in the corporate and project finance areas, covering advisory work, financial restructuring and fund raising for ventures in the People's Republic of China, Hong Kong and Macau. Dow MBf operates in the medium-size company range of the market-place and offers other financial services such as trade finance, foreign exchange, money market securities trading and private banking.

Conclusion

In his introduction to the 1985 Annual Report of Dow Financial Services, L. G. Merszei commented 'The international financial services market place is currently characterised by rapid change, deregulation and evolutionary developments not even dreamed of a decade ago. It is a challenging environment which has stimulated our growth. We are pressing forward in a well-planned, strategically conceptualised manner, providing our clientele with an increasingly broad range of specialised financial services through well-capitalised and very professionally staffed units in practically all major financial centers.' This is a fitting comment on the group's position today.

Above: L. G. Merszei, President and Chief Executive of Dow Financial Services/Royal Trust International

Kleinwort Benson: the development of a global securities business

Above a city merchants' tomb,
Was written RIP,
He sold a cargo CIF,
At prices FOB.

This little poem (attributed to Edward de Stein), a tombstone of a different sort to that prevalent in current markets, highlights what can happen to traders who have found it difficult to keep alert and abreast of the market.

A century or more ago the apocryphal cargo might have been tobacco shipped from Cuba by the Kleinworts or steel exported by the Bensons to the American mid-west where the railroads were fast being developed. The origins of the Kleinwort Benson group are steeped in nearly 200 years of financial and mercantile history and, although the original markets in which the founders of the firm were active may have been in physical goods rather than securities, the firm's success was based on trading expertise of the sort needed at least as much nowadays in the global securities markets as in the nineteenth-century trading markets.

Nowadays many institutions with little history behind them call themselves merchant banks, but the traditional British houses who first adopted the term did so for the good and simple reason that they were merchants who subsequently found it either more congenial or profitable, or both, to assume the mantle of bankers. It was a relatively easy transition, given Britain's dominant position in the financing of world trade and the consequential emergence of the famous 'Bill of London' which was the instrument that oiled the wheels of such trade.

The Kleinworts, Bensons and Lonsdales, whose names are reflected in the group holding company, all in fact had links with the grocery trade, although this common theme disappeared well before the merger of the group which today constitutes the largest independent British merchant bank in terms of published profits and balance sheet totals.

By the turn of this century Kleinworts had established the position of the largest accepting house, in terms of capital, in the City of London, and their reputation was such that when share issues for overseas household names like Woolworths, Continental Can and others were floated in the United States, Kleinworts were called upon as the leading overseas managing underwriter to help with the international distribution, even though its securities business was still only a small part of its overall trade.

For their part the Bensons were one of the founders of the investment trust movement and an important force in the financing of the US railroad system in the mid-nineteenth century.

Merger and expansion

The merging of these two houses in 1961 led to a period of sustained expansion which has resulted in a 100-fold increase in the balance sheet total over the last 25 years and a 35-fold increase in pre-tax profits.

This growth in business derived from all the main activities undertaken by merchant banks at this time such as commercial banking, investment management and corporate finance. But all the while, Kleinwort Benson, like the other merchant banks, was in the anomalous position of being unable to participate directly through the London Stock Exchange in its native stock market, while its international competitors in the field of corporate finance and capital markets from Continental Europe, the United States and the Far East were leading members of their own stock exchanges. Despite this constraint, the securities markets remained a central part of Kleinwort Benson's business, and indeed increased with every year that passed through the 1960s and 1970s, through the bank's prominent role as an underwriter of issues in the UK market, as an investor of its own capital, as a pioneer investment dealer on behalf of clients in the Eurobond market and as a manager of funds invested in every major securities market in the world, once more pioneering the move into emergent markets such as the Tokyo stock market in the 1960s.

During the latter stages of the first two decades of Kleinwort Benson's merged existence, the accelerating growth of the international capital market, the first signs of the wholesale securitisation of the banking business and Kleinwort Benson's ever-more prominent role in both domestic and international corporate finance led to an intensive internal review of the future shape of the world's financial markets and the implications for Kleinwort Benson's own operations.

Kleinwort Benson's new securities dealing room is one of London's largest, occupying 21,000 square feet

International focus

The referral of the London Stock Exchange under the Restrictive Practices Act may have taken place as early as 1976, but even by the beginning of the 1980s the prospects for real change still seemed remote and the world's two largest domestic capital markets, that of the United States and of Japan, seemed likely to offer more fruitful opportunities for immediate expansion, along with the ever-growing Eurobond market.

In the United States, Kleinwort Benson Incorporated had been participating as an underwriter of domestic US securities since 1969, but this activity had not been extended into any significant role in the secondary market of either fixed income or equity securities. In Japan, an early commitment to the future potential of this market, symbolised by the opening of the first representative office of any merchant bank in 1970, had within 10 years established Kleinwort Benson in an exceptional position in the Japanese corporate market, at just the time that these companies embarked on a massive programme of international capital raising, particularly through equity and equity-linked securities. In 1981, for instance, Kleinwort Benson managed more public international issues for Japanese companies than any other international bank and, of the Japanese domestic firms, only Nomura Securities managed more.

From the trading and investment standpoint, not only was the Eurobond market growing rapidly, but cross-border investment in domestic securities was accelerating. As Japanese institutions saw their foreign exchange restrictions lifted, significant parts of the United States deficit were financed by these institutions participating in the United States Government securities market, while in the US itself the previously parochial investment policies of the pension funds were giving way to a far more internationalist approach, with increasing funds being committed year by year to overseas equity markets, particularly those of the fastest growing economies in the Asia-Pacific region.

This strong international focus, which was applied by Kleinwort Benson to the planning of its own expansion in the securities markets from 1982 onwards, in the event coincided with the agreement between the Department of Trade and Industry in the United Kingdom and the London Stock Exchange in July 1983,

Above: the 23-storey head office building of the Kleinwort Benson group

whereby the Stock Exchange was exempted from the Restrictive Practices Act in exchange for the abolition of minimum commissions and a few other additional reforms. This enabled the bank to proceed during 1984 on its international approach to build a global securities business. Acquisitions in the United Kingdom were influenced by its international perspective and plans, and the acquisitions and expansion outside the UK were designed to enhance its capability in the UK securities markets of the future.

Even before Kleinwort Benson agreed terms with a UK Stock Exchange member firm for its essential entry to the UK domestic markets, it had simultaneously acquired the primary dealer in US Government securities in Chicago, ACLI Government Securities, and set up in Los Angeles a then unique company dedicated to the trading of US dollar interest rate swaps, a market which, since its inception in 1982, has rapidly become the dominant new factor in the continuing development and growth of the Eurobond market. Through its acquisition of ACLI Government Securities, which was immediately renamed Kleinwort Benson Government Securities, Kleinwort Benson became the first non-US bank to own a specialised primary dealer in US Government securities as an integral part of its international securities and capital markets operations. The central role of the US Government securities market within the US dollar market generally and, as a result, within the world financial markets as a whole, made a major presence in this market an objective of the highest importance in developing all other securities' activities within the group. The size, sophistication and liquidity of the US Government securities market makes it inevitably the breeding ground for the most advanced trading technology and techniques, with all the consequential benefits to the group's operations in the UK Gilt market, the

Eurobond market and other fixed income markets around the world.

During 1984 a further acquisition was completed in the United States, of a specialist financial futures broker, Virginia Trading Corporation, with an outstanding reputation for both research and execution in the important financial futures market in Chicago.

The combination of these newly acquired operations gave Kleinwort Benson a capability in the US dollar interest rate markets which is unequalled by any other European bank and acts as a linchpin for all the bank's trading in debt securities.

The choice of Grieveson Grant as Kleinwort Benson's partner in the UK securities markets of the future was based on similar analysis and priorities. Grieveson Grant was founded in 1875 and had, by the early 1960s, established an outstanding position in the UK Gilt market as one of the seven firms which dominated the execution of transactions in the gilt market throughout this time. This base, augmented by the acquisition by Kleinwort Benson of one of the specialised gilt jobbing (market making) firms, Charlesworth, shortly after that of Grieveson Grant, was designed to allow the group to maintain a major market share of the reformed and deregulated gilt market among domestic investors and to develop further the outstanding international business begun by Grieveson Grant – particularly with Japanese institutions – alongside the group's US Treasury and Eurobond activities.

Grieveson Grant also had a strong and growing share of the UK equity market but, as important, had once more developed an international approach to all aspects of the equity business: international investment demand for UK equities was developed (in the United States, Japan and Continental Europe) with as high a priority as the expansion of the domestic investor base, while business in overseas equities markets was expanded with equal vigour. The firm's research in Continental European equity markets – the Cinderella of the 1970s but a major growth market in the 1980s – is widely acknowledged as being among the best in the world (as a *Euromoney Corporate Finance* survey showed), while rapid growth was also achieved in the US equity markets – with membership of the mid-western US stock exchange having been acquired in 1975 – as well as in the Japanese equity market.

The attraction of Grieveson Grant to Kleinwort Benson was based on its combination of major domestic market share with a significant and balanced international business. In addition, an outstandingly successful investment management business, both for institutional and private clients, provided a basic support for the business which, when merged with Kleinwort Benson's own investment management operations, would produce one of the largest investment management operations in London.

Even though the rules of the London Stock Exchange did not allow Kleinwort Benson to take full ownership of Grieveson Grant (or the gilt jobbing firm of Charlesworth) for nearly two years after the initial agreement, the international operations of the Kleinwort Benson Group were able to draw on the resources of Grieveson Grant in their expansion even before the final completion of the acquisition. An application by Kleinwort Benson to the Japanese Ministry of Finance for securities licences had pre-dated the reaching of agreement with Grieveson Grant, but the ability to incorporate the employees of Grieveson Grant's representative office within Kleinwort Benson's own Japanese operations (which were awarded all four relevant securities licences in November 1985) enabled the Tokyo securities branch of Kleinwort Benson International to commence operations immediately. It thus enjoyed from the start a strong and broadly based business in the research and sales of Japanese equities to both international and Japanese domestic investors and in the trading and sales of US Treasuries gilts, yen government bonds and Eurobonds to the huge pool of Japanese international liquidity. In addition, the initially small but growing appetite by the same Japanese institutional investors for overseas equities has given rise to a significant volume of business in UK equities as well as in Continental European and US stocks.

In the United States, Kleinwort Benson Incorporated was able to expand its traditional role of underwriting domestic US equity offerings to being an active trader in the American Depository Receipt (ADR) market for international equities, where Grieveson Grant's research and reputation – for Continental European equities in particular – was a vital factor in the establishment of this unit.

Although the US, Japanese and Eurobond markets are very much the world's largest securities markets, and therefore the highest priority for the Kleinwort Benson group alongside the domestic UK market, some of the second-tier capital and securities markets have been targeted. In Australia, for instance, a microcosm of Kleinwort Benson's global approach was developed through Kleinwort Benson Australia taking significant shareholdings in a leading Australian stockbroking firm, Hattersley, Maxwell, in one of the top three traders in long-term Australian Government bonds, the Australian Gilt Company, and in a leading trader in short-term Australian Government paper, the Colonial Mutual Discount Company.

The markets of the future

The challenge confronting those firms aspiring to be successful and profitable participants in the securities markets of the future is not merely to survive the consequences of the UK market's deregulation or Big Bang, but to combine the quality of service traditionally associated with smaller specialised houses with the trading capability, information services technology, capital and all the other resources that only a large and firmly established group can achieve.

Every successful bank participating in the global securities markets undoubtedly requires a strong home base, reflecting a leading position and prominent market share in its domestic securities market; but in an open and liberalised market such as the UK will have at the end of 1986, a parochial approach is unlikely to enable a UK firm, however focused and specialised, to maintain even its domestic market share as competition from both other domestic and foreign houses increases. Kleinwort Benson, historically and philosophically dedicated to an internationalist rather than a parochial approach, has responded to the deregulation of the UK capital market with a commitment to all the major securities markets of the world.

Left: the sale of 50.2 per cent of British Telecom in November 1984, in which Kleinwort Benson acted for HM Government, was the largest issue of shares that the world had yet seen, raising just under £4 billion.

Macfarlanes' offices at 10 Norwich Street, specially designed for the firm

Macfarlanes: ready for the future

Like most revolutions, the present changes in the City will make a continuing combination of demands on its participants, a mixture of the stimulating and the sobering: tremendous commitment and capital for investment in the best people, premises and technology; adaptation to new pressures and competition from major overseas institutions; the disappearance of historic and comfortable cartels; and a necessary and keener appreciation of the dynamics of world markets and practices.

There was no possibility of escape from all this, and little promise of successful half measures. Those in the City with the necessary vitality and pragmatism have long since made such preparations for Big Bang as would enable them to serve their clients – and, it must properly be said, their own subsequent self-interest – as formidably well in the future as they had in the past.

The best firms of solicitors in the City, like their counterparts in other professions, have been growing apace not simply because they are able, diligent and virtuous, but also because they have not been complacent. They have for years observed and fostered the new skills and attitudes found necessary by the market-places in which they function. This is certainly true of Macfarlanes: we have for over 100 years sought to apply distinct qualities of judgement, commitment and effectiveness to the affairs of our clients – and whenever possible, some good humour as well.

A strong history

Macfarlanes was founded in 1875, initially concentrating its practice on service to many of the wealthy and land-owning families of England and to the trusts which they created. From the outset the firm provided sophisticated legal and fiscal advice for the preservation and safe increase of that wealth.

From such beginnings, our practice has seen sustained and wide-ranging growth, so that the greater part of the work now undertaken is for our corporate and institutional clients. However, Macfarlanes still devotes a larger proportion of its resources than other comparable City firms to the interests of individuals and to their financial affairs.

Since 1960 Macfarlanes has trebled in size and has a current staff of 260, including some 115 lawyers. At the moment there are 27 partners in the firm, whose average age is 42. We are an ambitious and efficient firm, modern in outlook and technique. Growth has been carefully managed; Macfarlanes has never merged with another firm and, although we anticipate continuing strong growth in the next few years, we have avoided size as an end in itself.

Macfarlanes has always been based in the City of London, and in 1983 we moved to a new building at 10 Norwich Street which was specifically designed for the firm and its practice. These modern offices provide excellent facilities for the practice of law, including support services based on the latest available communications technology. A powerful new IBM integrated computer system has just been installed for our data and word processing, accounting and document retrieval systems.

Growing with the City

The firm's respect for its own traditions of excellence and the traditions of the City are fundamental to Macfarlanes' continued growth and success. London's development as a major financial and international business centre has historically provided extraordinary opportunities for the firm to acquire new expertise and clients; hence our conviction that the structural changes in the City will provide a basis for continuing expansion and creativity in our main areas of practice. For example, the property department is advising the emerging financial conglomerates on the redevelopment and occupation of office buildings which are now being constructed on an entirely new scale; individuals who have benefited in various ways unexpectedly and greatly from the acquisition of their businesses have looked to us for tax and financial planning advice; and the new environment is making fresh demands on the skills of our litigators.

However, the reader will be mainly concerned with the impact Big Bang is having on our company and commercial law department. This department deals with the whole area of corporate and commercial law from public company take-overs to one-man business start-ups. Clients range from multinational corporations to small domestic businesses and include merchant banks, stockbrokers, investment institutions and a variety of manufacturing and transport companies, specialist electronic firms, advertising agencies, insurance companies, retail and franchise organisations and publishers, as well as individual entrepreneurs and investors.

Within this wide range of clients, it is appropriate, in the context of Big Bang, to emphasise our role in representing banks, brokers and other financial institutions for whom the City is 'home' and companies who invite us to represent them in their dealings with the City institutions. As the regulatory framework becomes more complicated and competition within the industrial and financial market-place increases, the role of the lawyer is becoming central in transactions that formerly were almost the exclusive preserve of the financial houses. Evidence can be found in the increasing use of the Courts in hostile bids and in challenges to the authority of the Office of Fair Trading. Precedent – the solicitor's touchstone – suggests that this process is unlikely to be reversed.

There are many factual indications of the sharp and continuing growth of our corporate practice. Over one-third of our partners are now committed to corporate work; we are among the 10 firms in England which have acted most frequently as solicitors to the sponsoring broker or banker in USM issues; we are being instructed in an increasing number of takeovers and have acted, most often for the offerors, in eight bids in the first six months of 1986 alone; the market capitalisation of the listed companies with which we are publicly associated now totals many billions of pounds; and our specialists in such fields as venture capital, management buy-outs and offshore investment funds are particularly active.

Two special facets

Underlying these statistics are two facets of our practice which bear special discussion. Both are at least partly a function of the great changes taking place; both will become increasingly important and valuable as those changes are fully implemented and developed: they are the breakdown of traditional professional allegiances and the City's ever increasing internationalism.

The City has for many decades been an exceptionally traditional market and something of a closed circle. That indeed has been one of its great strengths: the bankers, brokers and others of real influence were relatively few, and all knew, or knew of each other, very well. They had, as part of their properly conservative practices, predominantly used the services of a small number of professional advisers. While true of other professions also, it was particularly true of solicitors. Almost all of the largest firms of the present day were already large firms, by the standards of *those* days, between the wars or soon after World War II. They had then a virtual monopoly of the most important banking and corporate work, and the volume and scope of that work was never so great as

Some of the partners in Macfarlanes' Company, Commercial and Banking Department

to make conflicts of interest a real problem.

The last few years have seen an erosion not only of those barriers but also of the attitudes that first created them. Both barriers and attitudes are now, for many and certainly for Macfarlanes, in a state of welcome disintegration. Many of the newer and most powerful forces in the City, whether institutional or industrial, are not simply prepared, but positively willing, to employ professional firms that would once have been on the periphery of major City work. Those firms must, of course, have the fundamental qualities and skills which make them fit to compete on equal terms with anyone; but the broader horizons, keener criteria and fresh imagination that are now permeating the City and its practices provide a splendid stimulus to firms like ours.

We have made our preparations to take advantage of these opportunities in various ways: recruitment of an increasing number of articled clerks from the universities and of solicitors from elsewhere; the best possible working practices in modern premises; the latest technology manned at week-ends and well into the night as necessary; continuing training and retraining at all levels in new aspects of practice; and a spirit of commitment and cohesion throughout the firm. So it is that we now act for that wide range of corporate clients alluded to earlier, many of them new, some already household names and others with great emerging potential.

The other important facet reflects on our long-standing international outlook. For a decade or more we have forecast and prepared for the disappearance of frontiers, whether of culture, distance or time. We have developed contacts with lawyers and law firms throughout the world, particularly in America, Europe and the Far East. We have monitored and recorded their performance carefully for the benefit of our clients. Lawyers from America, Japan, Holland, Italy, Switzerland, Australia and elsewhere have recently trained and worked with us. We, in turn, send lawyers from our firm on secondment overseas: in 1986 alone to firms in Miami, Boston and New York. Our international contacts have most naturally been fostered by the continuous flow of work with a foreign element. We see many signs of this flow increasing in parallel with the burgeoning overseas presence in every aspect of City activity, and the increasing expatriation of City skills abroad.

A substantial and constructive part

Macfarlanes' existence in the City for over 100 years has drawn heavily on the traditional City virtues of honesty, trust and the appreciation of real worth. These virtues will become increasingly important in the volatile environment of Big Bang. In the context of them we have in these notes briefly marked our readiness to participate in all that is happening in the City; the preparations, both in materials and perspective, that we have made to do so; and the belief that Macfarlanes' experience and skills, applied forcefully, functionally and distinctively on our clients' behalf, will play a substantial and constructive part.

Midland Montagu, Samuel Montagu and Greenwell Montagu: Midland Group's response to Big Bang

The new building at Billingsgate, shortly to be occupied by personnel from Samuel Montagu, Midland Group Treasury and Greenwell Montagu. Inset is Samuel Montagu's present head office

The Midland Group's response to the challenges presented by Big Bang has been emphatic and wide ranging. As a broadly based financial institution, the Midland Group has always recognised the importance of providing an organisation which meets the needs of its clients and customers in the UK and worldwide. For this reason, it has created a structure based on four broad business sectors: retail, corporate, international and investment banking.

The key feature which distinguishes Midland Montagu's investment banking activities from those of the other merchant and investment banking groups is the nature and extent of the resources which have been devoted to this sector.

The component elements of the sector comprise a synergistic blend of traditional expertise and ultra-modern financial skills which extend across the capital markets, the money markets and the credit markets. Midland's Treasury operations have always demonstrated an ability to anticipate market developments; Midland was the first to merge its domestic and international treasury operations and more recently Midland Treasury has been at the forefront of technological sophistication. Samuel Montagu has a history of market making and trading which goes back over 130 years, while W. Greenwell has been active in the equity and government securities markets for almost 120 years.

Midland Group Treasury

In addition to the longstanding activities centred on the foreign exchange and money markets, Group Treasury offers recently introduced products such as swaps, options and futures as well as state-of-the-art synthetic products which can only be created through highly complex techniques.

The inclusion of Midland Treasury in the investment banking sector is a key element of the Midland Group's global investment banking strategy. Midland Treasury funds and manages an asset book in excess of £56 billion; this generates naturally occurring cash flows and maturity positions, which, combined with the use of sophisticated financial techniques, gives the ability to position interest rate and currency swaps and create innovative and effective financial products.

Samuel Montagu

Samuel Montagu was established in 1853, with the release of the following announcement: 'Mr Edwin L. Samuel, Banker and Bullion Merchant and Mr Samuel Montagu have established a Bullion and Foreign Banking Office. The highest prices allowed for every description of Gold and Silver in Coin, Bars, Plate, Lace etc. Platina and Palladium bought. Foreign Bank Notes and Coupons of every country exchanged at the most favourable rates, Drafts granted, Bills collected, and money forwarded to all parts.' Some of those services are still very significant for the bank today, and over the years many more have been added in order to service clients in the UK and internationally, from offices in over 20 countries.

Samuel Montagu's wide range of activities is reflected in its membership of the following bodies: The Accepting Houses Committee, The Issuing Houses Association, The London Gold and Silver Markets, The London Gold Futures Market, The London International Financial Futures Exchange, The International Primary Market Association, The Association of International Bond Dealers and The Euronote Association.

Following the relaxation of exchange restrictions and the reopening of the London Gold Market in 1954, Samuel Montagu was able to consolidate its position in foreign exchange and bullion dealing, and to expand its business in acceptances and commercial credits, sterling deposits, investment advice, portfolio management, capital issues and corporate finance.

The 1960s marked one of the most significant developments in Samuel Montagu's history, namely, its involvement in the Euromarkets. As far back as 1927 Samuel Montagu had participated in major capital market operations, these had largely been sterling related. The opening year of the bank's role in the Eurobond markets was 1963, when Samuel Montagu sponsored and completed the first dollar loan to be made to a borrower since World War II. The loan for the Kingdom of Belgium has a claim to be considered as the first ever Eurodollar bond.

For some time following the Kingdom of Belgium issue, Samuel Montagu focused its efforts on secondary rather than primary market operations. However, the early 1980s marked Montagu's return to primary market activities with the recruitment of David Potter, who set up the International Capital Markets Division. Since 1982 Montagu has consistently been one of the top managers of both domestic sterling and Eurosterling issues, and in 1985 Samuel Montagu lead-managed and co-managed some 233 international bond issues totalling US$40 billion. Montagu's strength in sterling was further reinforced in 1985 with lead-managements in Bulldog, Fixed Rate Eurosterling and sterling FRN issues.

Greenwell Montagu

Greenwell Montagu was founded as W. Greenwell & Co by Walpole Greenwell in 1868. Today it occupies a coveted position as one of the country's leading stockbrokers, offering a comprehensive range of services to institutions, companies and private investors in the fields traditionally associated with the Stock Exchange.

Greenwell Montagu's dominant position in the gilt-edged market has foundations in respected research of monetary economics. In Gordon Pepper, Greenwell Montagu has one of the most renowned market commentators. The gilt operation currently uses the most modern techniques for both communication and calculation while the Greenwell Montagu price service on the Stock Exchange TOPIC television service is one of the most accessed in the market.

Greenwell Montagu has a strong presence in the UK equity market providing a sales, distribution and research service to most major UK investment institutions, as well as many abroad, especially in Europe and the United States. Through its extensive research capability it publishes regular commentaries on industries and companies as well as views on the market and economy. The sales desk works with the corporate finance department in underwriting issues and placing lines of stock.

Greenwell Montagu's corporate finance department acts as a corporate broker to some 100 quoted companies. It is a link between the Stock Exchange and the company, providing advice on the raising of capital as well as keeping the company's management in touch with economic and market developments. It also has a strong and established reputation in the field of Bulldog issues, the raising of sterling finance for foreign governments.

The private clients department provides highly specialist and specifically tailored services for private individuals and trusts, together with bank trust companies, bank branches, accountants and pension funds. All aspects of asset management are examined including traded options, gold coins, unit trusts, and certain non-Stock Exchange transactions. The taxation implications of investment policy are fully considered and taxation planning is available.

Smith Keen Cutler, the Birmingham-based stockbroker, acquired by Greenwell Montagu in April 1986, was founded in the 1840s by John Smith who played a key role in establishing the Birmingham Stock Exchange in 1845.

Smith Keen Cutler is now a leading provincial broker, with an extensive private client and retail distribution network. Earlier this year it opened the first share shop in a high-street branch of Midland Bank.

The Midland Group's investment banking activities

The restoration of Midland Bank's 100 per cent ownership of Montagu in 1985 laid the foundation for the development of a global strategy for the combined investment banking activities of Montagu and Midland. Thus, following the completion of Montagu's purchase of W. Greenwell & Co, Midland announced its decision to concentrate all its investment banking activities into one sector. This involved bringing together Midland's Treasury operations headed by Ernst Brutsche, Midland's investment-related activities, the merchant banking activities of Samuel Montagu and the Stock Exchange business of Greenwell Montagu. To this organisation has been added the commitment to enable the Midland Group to be at the forefront of the new markets as they develop.

Midland's investment banking capabilities are represented by a combination of financial resources, experience and expertise which is virtually unrivalled in the sterling markets. The reality of Midland's investment banking presence is embodied in the new building on Lower Thames Street, which will be occupied late in 1986. Here the group will have over 750 dealing positions and, under one roof, the ability to trade in the full range of financial instruments.

For both Greenwell Montagu and Samuel Montagu to continue to compete successfully in the deregulated markets after Big Bang, access to a large capital base is clearly essential; this is provided by the Midland Bank, which has a capital base in excess of £4 billion.

Complex risks involved in trading and placing the huge variety of different securities mean that technology is now a major consideration. Robert Logan, Chairman and Group Chief Executive of Samuel Montagu, strongly believes that effective application of technology in the investment banking business will be critical to the sector's future success. In view of this Samuel Montagu has invested in Electronic Banking Systems Limited (EBS). EBS is developing efficient dedicated systems and communication capabilities necessary to support the implementation of corporate strategy.

Market making

Greenwell Montagu's experience in the gilt-edged and equity markets will be complemented by Samuel Montagu's long-established market making expertise in bullion, foreign exchange and Eurobonds. Together the group will be positioned to capitalise on the opportunities in this area. Post Big Bang, Greenwell Montagu Gilt Edged will be one of the gilt-edged market makers and Greenwell Montagu will be a market maker in leading UK equities.

Ernst W. Brutsche, Chief Executive (Group Treasury), Midland Group

Distribution

An extensive distribution network will be another characteristic of the successful investment bank. Greenwell Montagu with established distribution capability in gilts, UK equities and domestic sterling debt instruments, such as the recently established commercial paper market, will continue to provide the highest quality service to institutional clients. Distribution to clients outside the major institutions is also considered important and here the Midland Bank's branch network provides access to numerous small investors.

Research

Outstanding research capability is another feature of Greenwell Montagu. Research is a proven determinant of market share, and expertise in this area will be an invaluable tool post Big Bang. The Greenwell gilt-edged research team monitors the macro-economic trends and its team of economists, led by Gordon Pepper, writes regularly on the gilt-edged market and associated topics in the *Monetary Bulletin* – one of the most respected of all stockbrokers' publications.

Greenwell Montagu equities has one of the strongest research departments in the City, providing company and market analysis. It has rated analysts in banking, chemicals, financial services, insurance, pharmaceuticals and property, as well as covering all the other major companies in the United Kingdom. The wealth of knowledge in the company's research department brings great benefit to the Group as a whole. The knowledge of UK companies is now being expanded to provide research on European companies. The links with Samuel Montagu and Midland Bank in Europe provide an excellent opportunity for developing research which in turn will lead to sales and trading opportunities. Additionally, research on the Far Eastern markets is now being originated in Hong Kong and distributed and sold in Europe.

Investment management

The Midland Group has moved quickly to fill the gap in its investment management capability arising from the sale of Montagu Investment Management. Montagu Investment Limited manages institutional and private client funds on a discretionary or advisory basis. In addition, Montagu Investment Limited provides unit trust management services and equity finance.

The global investment bank

For the Midland Group, Big Bang has been only one of a number of important developments in the capital markets around the world. The traditional role of banks as lenders has been changed by securitisation, and Midland has responded by applying resources to the development of its network in over 40 countries to build a global securities business in all major currencies and financial instruments. The investment banking sector will be using subsidiaries and associates in the Far East, the United States and Europe to serve clients in their domestic markets as well as the international markets based in London, Tokyo and New York.

Right above: Robert Logan. Right: Christopher Sheridan

Price Waterhouse's all-round service to market players

The international financial services industry is going through a period of fundamental and accelerating change. New techniques and new instruments are proliferating in response to large and unpredictable variations in inflation, interest and exchange rates. Financial institutions and their customers alike are becoming more sophisticated in their approach to the capital markets and the provision of financial services. Competition between financial institutions is increasing as a result of these developments and in response to changes in the regulatory framework.

In anticipation of the increasing demand for professional services to cope with the changes in the securities industry, Price Waterhouse has been steadily building up its expertise, experience and resources and now offers a comprehensive range of services to clients throughout the world.

Price Waterhouse's banking and financial service clients range from major world organisations through a significant proportion of the top 100 banks in the world to many smaller institutions. Each client has its very particular needs which receive independent and practical advice from the engagement partner and his multi-disciplinary team.

Some of Price Waterhouse's recent engagements in support of market participants have been:

- the development of a comprehensive set of risk control procedures for a major Eurobond trader;
- project management of the development of a major market making support system for a consortium of five major securities houses;
- advising on the selection of computer systems in a regulatory organisation which may form the basis for a recognised exchange;
- assistance with design of sophisticated limits systems and comprehensive control procedures for an international bank involved in the options markets;
- the establishment of capital markets and gilt market making operations for several leading banks, including taxation advice and system selection and implementation;
- advice on the development of the systems capability for a major international bank to support international equities trading across the triad of London, New York and Tokyo; and
- advice on product positioning and delivery locations of a major international clearing house

In particular, Price Waterhouse is closely involved in helping many banking and financial institutions to prepare for the Big Bang in London and cope with worldwide developments in capital markets.

History

Price Waterhouse has always been heavily involved with clients in the financial services area. The firm was founded in 1849 by Samuel Lowell Price, who practiced as an accountant from Gresham Street in the City of London. On 1 May 1865 Samuel Price was joined in

Price Waterhouse's office in Frederick's Place in the City of London in 1899

partnership by Edwin Waterhouse and William Hopkins Holyland. The principal work undertaken by the partnership in its early years was that of liquidations, bankruptcies and arbitrations. The granting of limited liability to joint stock companies in 1862 gave further scope for accountants as this led to the creation of the role of official liquidator. The firm was heavily involved in liquidations and in 1866 was appointed to wind up an international bank. The Limited Liability Acts led to several of the large joint stock banks considering the possibility of audits. Price Waterhouse were soon appointed auditors to many of the major banks, including the National Provincial Bank of England and the London and Westminster Bank. In 1890 Mr Lloyd of Lloyds Bank first sought the services of the firm.

Banks also began circulating audited statements of affairs to their customers which was followed by a series of amalgamations between both private and joint stock banks. Price Waterhouse was heavily involved both in negotiating the terms of amalgamation, and also in auditing these larger institutions. In 1895 Edwin Waterhouse was involved in the amalgamation of Gurney & Co of Norwich with Barclay & Co which later became Barclays Bank plc.

Price Waterhouse was also innovative in other countries. The St Louis office opened in November 1901 and prepared in 1902 the first audited financial statements of the Mechanics National Bank of St Louis, and the Mississippi Valley Trust Company. The importance of audited figures was noted by the *St Louis Republic* who reported that 'The Mechanics Bank is the only national bank in the United States which gives its stockholders and patrons the advantage of a disinterested examination by chartered accountants.'

Price Waterhouse soon grew in size. In 1900 there were 86 partners and staff in London and 27 staff in the United States. Price Waterhouse had opened offices in New York (1890) and Chicago (1892) as these cities grew into the two financial centres of America. Worldwide development soon followed with offices opening in Canada (1907), which in 1910 examined the accounts of three major banks, Buenos Aires (1912), Rio de Janeiro (1915), St Petersburg (1916), Paris (1917) and Rotterdam (1919).

Price Waterhouse firms worldwide now employ over 30,000 people in 378 offices providing services to clients in 98 countries. The worldwide expansion has closely followed the international banking market and services have developed out of their close association with the financial communities in the major financial centres of the world. The financial sector clients served by Price Waterhouse include some of the largest multinational institutions and a wide number of local ones. This strength in financial clients is worldwide but it is particularly marked in New York, London and Tokyo. The lead banking specialist partners in these cities – Tom Macey, Nigel Buchanan and Toshifumi Nagasaka – work very closely together to provide a coordinated multidisciplined service for their clients.

Price Waterhouse had its foundations built on solving banking problems. The firm has been heavily involved in investigating the banking collapses of the 1930s, 1970s and 1980s and in recent years has advised banks on the viability of many international companies. Over the years the range of services offered has increased dramatically.

The banking and financial services specialism within Price Waterhouse

From the earliest years Price Waterhouse has been committed to a programme of industry specialism for their staff. In 1910, Sir Arthur Dickinson took an entire audit team from New York to examine the Bank of Nova Scotia main office in Toronto. This ensured that the staff involved in the examination were all experienced in banking. The banking specialism which was established very early in the firm's evolution continues to be at the forefront of its development in this area.

In order to ensure that the banking industry specialists are kept informed of key business and technical developments within the industry, the firm has established a central support specialist group whose responsibility it is to coordinate the firm's services to the banking community and to ensure effective communication between the banking specialist staff. In addition, a new products team has responsibility for examining the business, control and accounting implications of new banking products as they are developed.

It is Price Waterhouse policy to allow staff to specialise in banking and financial services at an early stage in their careers. This policy not only ensures that by the time the staff become partners and managers they have considerable experience of the banking and financial services industries but also ensures that the field staff have banking experience and are familiar with the industry. Price Waterhouse believe that this policy allows it to provide a unique service to its banking clients by providing them with highly experienced teams.

Southwark Towers – Price Waterhouse's present office in London

The multidisciplined banking and financial services team in the UK is coordinated by Nigel Buchanan and is split into various specialist areas which include the following:

Capital markets and dealing – Richard Kilsby
Securities – Leslie Cousins and Andrew Coleman
Investment management – Steve Dent
Broking – John Harley
Organisation, strategy, management information and risk management – Mark Austen and David Roe
Supervision – David Crowther
Retail banking – Richard Rimmer
Building societies – Roger Hughes
Leasing and instalment credit – Richard Stevens
Treasury management – Paul Reyniers
Insolvency – Mark Homan and Colin Bird
Viability studies – John Heywood and Mike Gercke
Commodities – Desmond McCann
Taxation – Philip Gillett
Venture capital – Michael Langdon
International trade (international taxes, sales taxes and customs duties) – Mike Maskall and Deirdrie Davis
Information technology – Bob Pamplin and Alan Johnson

All these and other partners may be contacted in London on 01-407 8989 or via your local Price Waterhouse office.

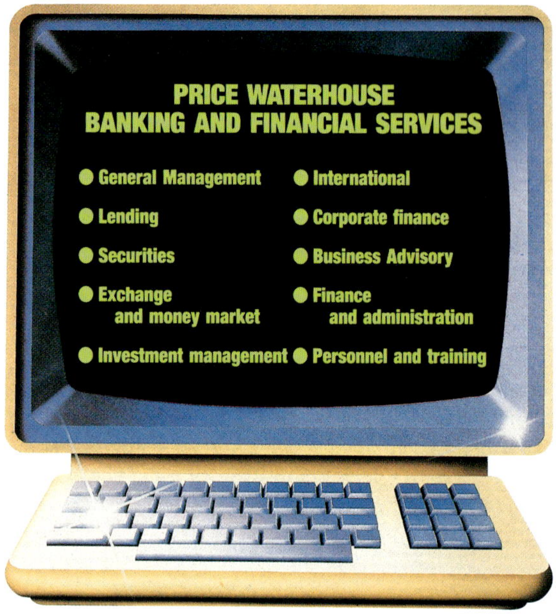

Price Waterhouse services

Price Waterhouse has adopted a multidisciplined approach to servicing its banking and financial sector clients. Therefore, each task or project can be performed by staff selected from a variety of disciplines. Price Waterhouse provides services to the following areas in a bank or financial institution:

General management

The senior management of a financial services organisation is faced today with ever increasing challenges as a result of the changes taking place in the financial services sector. Day-to-day operations are becoming more complex as new financial markets and financing methods are developed. At the same time the business climate is becoming more competitive as other institutions broaden their product base. These developments have resulted in a need for senior management to concentrate upon risk management.

Price Waterhouse can advise management on the preparation of strategic plans, including the identification of suitable areas of business growth, developing and establishing effective plans and policies to exploit those areas and on the management of risks and cost reduction.

Lending

As the range of lending instruments expands, both borrowers and lenders have to make decisions on financing, bearing in mind conditions in a variety of currency and capital markets. Price Waterhouse perform loan assessments, feasibility studies, business reviews and special audits, and advise on the structuring of financing deals and the planning, coordination, control and supervision of capital projects.

Price Waterhouse has considerable expertise in insolvency practice and procedures in many parts of the world. The work in this area can include assistance to borrowers to demonstrate and improve their ability to repay debt. Price Waterhouse can also act as receivers for banks with floating charges who wish to realise their security and will generally try to continue the business for sale as a going concern.

Securities

In order to survive and succeed, given the major changes in the various securities markets and the development of new markets and instruments, organisations in the industry need to reassess their business strategies in relation to the changing environment. In developing a strategy organisations will need to recognise their strengths and weaknesses and identify their critical factors for succeeding in their chosen market niche. In implementing a strategy, organisations may consider possible mergers and acquisitions, examine their personnel and technological resources, and effect changes to match their revised objectives.

One of the significant forces of change is the use of new Information Technology (IT). Price Waterhouse has both the experience and the resources to assist financial institutions both to develop an appropriate IT strategy and to implement the agreed proposals. Within the broad range of IT services, Price Waterhouse has developed particular expertise in a number of areas of special relevance to the securities industry such as dealer support systems, treasury management systems, portfolio management systems, research analyst support systems, and external information services.

Day-to-day operations cannot be performed effectively without the use of increasingly sophisticated technology. With this increased use comes a greater exposure to the risk of fraud or error and a high degree of dependence on the availability of systems to operate the business. Data security and control is therefore a key facet in the securities industry and Price Waterhouse have had considerable experience in advising on the adequacy of security and control features and in assisting in implementing procedures for clients in the securities and financial business sectors.

Exchange and money market

A fundamental part of a financial institution's business is its activity in the various foreign exchange and related money markets, both on behalf of its clients and on its own account. Although good performance can make a major contribution to financial results, control breakdowns can represent substantial risks. Price Waterhouse brings to this area a wide range of experience in assisting its clients to control, manage and report on these activities.

Price Waterhouse has extensive experience in assisting clients to design and introduce systems and procedures for the recording, reporting and control of their activities in foreign exchange, futures and commodity markets as well as other money markets such as the interbank and Eurocurrency markets. This assistance includes advice on the evaluation, selection, development and implementation of the full range of computer systems support and back office procedures.

International activities

Trade is international and banking follows trade. Banks now habitually operate throughout the world and provide services which are not confined by national boundaries. In recent years the international activities of many financial institutions have grown dramatically and they now look to their international business to provide a significant proportion of total profits.

Price Waterhouse provides its services through a worldwide network of offices and is a prominent member of the financial community in the countries in which it is represented. Clients are therefore offered assistance and advice based on a thorough knowledge and experience of local conditions. In the major financial centres, Price Waterhouse has accumulated considerable experience in helping banks in setting up operations and has assisted with many aspects of such projects, including feasibility work, project management, systems development and review of tax implications.

Investment management

Institutions which aim to offer a full range of financial services increasingly wish to provide a comprehensive investment management service based on an expert knowledge of securities markets. This expertise must also be applied to an institution's own investment portfolio. Recent years have seen significant developments in the investment market, with the growing internationalisation of portfolios and the diversification of investment instruments. This increasing sophistication, together with a growing competition for funds, has required corresponding progress operationally, most notably in the use of computer systems and advanced telecommunication networks. Price Waterhouse provides a range of services to its clients with investment management activities in such areas as systems support (including telecommunications), taxation advice, evaluation techniques, audits and viability studies.

Corporate finance

An increasing number of banks and other financial institutions are offering specialised assistance to companies seeking growth by acquisition or merger and/or considering quotations on stock exchanges or other equity markets. Price Waterhouse offers services which assist the corporate finance departments in providing this support to their clients. Price Waterhouse regularly acts as reporting accountants to companies seeking listing of their shares on a stock exchange. This generally involves the preparation of detailed reports for the directors and sponsors.

Business advisory services

The financial community is increasingly involved in advising smaller businesses. Price Waterhouse can assist this advisory function by training business advisory staff in such fields as accounting and information systems, costing, stock control and data processing concepts. In addition, Price Waterhouse can support business advisory staff in market and feasibility studies, systems development, taxation and treasury management.

Finance and administration

Price Waterhouse provides a comprehensive range of services to the financial and administrative management of financial institutions. These functions have been expanded and developed in recent years to encompass audit and accounting, profitability reporting, taxation, accounting and information systems, efficiency and automation, and computing.

Personnel and training

More than ever commercial success depends on an ability both to attract and retain sufficient high calibre staff. Banks need to provide effective personal and career development for all their staff to meet their individual needs and to provide the human resources necessary if demanding commercial objectives are to be met. Price Waterhouse services in this area include executive selection, advice on remuneration packages and all aspects of training and career development.

Big Bang, the rapid development and the increasing internationalisation of the financial markets place exceptional pressures on management and lead to significant opportunities and risks. Management, therefore, need an experienced and multidisciplined team to advise on the risks involved in the market and on the systems necessary to take advantage of the opportunities. Price Waterhouse is uniquely placed to provide independent and constructive advice.

Shearson Lehman

– when know-how tells

At no time in the history of the City of London has so much time, effort and money been put into, what is for most, still a huge questionmark. Whoever first dubbed deregulation of the London equity markets 'Big Bang' was surely anticipating an almighty explosion. Will he be right? Or was the wife of one prominent City banker, bemoaning the absence of her husband for nights on end, nearer the mark when she was overheard at a cocktail party saying 'Big Bang? It seems more like a slow puncture to me.'

There is no guarantee that the same effects arising from the abolition of fixed commissions on the New York Stock Exchange in May 1975 will be felt in London, but there is a good chance that they give a foretaste of things to come. Eleven years on the results speak for themselves. By the first quarter of 1986 the top seven investment houses in the United States – Shearson Lehman included – had 91.5 per cent of the primary debt market. As François de St Phalle, Senior Executive Vice President of Shearson Lehman in the US says, 'It will be difficult for many of the medium-sized banks to compete. If, for instance, they focus their capital on the stock and bond markets, they may have to constrain other activities in, say, swaps, international sales, yen trading or building infrastructures.'

Despite all that has been said, written and hypothesised by the banks, little has come from the investor. But it is he that should see the benefit from the vast sums of capital being committed. Not only is it likely that commissions and overall dealing costs will be reduced, but also stock and bond quotations will most probably be

Above: Jacques Gelardin, Chairman of Shearson Lehman Brothers International
Right: Bill Harrison, head of Investment Banking in the UK at Shearson Lehman Brothers International

priced on finer spreads if the experience of the New York Stock Exchange is anything to go by. Since May 1975, quotations with spreads of ¼ point or less have risen from 44.9 per cent of all quotations to 60.7 per cent and at the same time liquidity in the market has remained good. This augurs well for investors in the London market.

Among the US investment banks operating in London, Shearson Lehman alone has made the commitment of acquiring a UK stockbroking firm; since April 1986 it has owned 100 per cent of L. Messel, a century-old stockbroker which was one of the largest brokers in gilt-edged securities and UK equities. Their strong research in British and foreign stocks, their connections with large UK companies, considerable experience in new issues, the gilt-edged market and financial futures are an excellent fit for Shearson Lehman which is one of the top investment banks in the world.

As Jacques Gelardin, Chairman of Shearson Lehman Brothers International, puts it, 'We are an extremely well balanced firm combining a very large capital base with traditional banking experience and a powerful worldwide trading and distribution network. We are positioning ourselves in what we think will be the next phase in the development of the international capital markets.'

'The purchase of Messel consolidates our existing position. We are already involved in those businesses which we see as growth areas for us – notably gilts and equities – but we would not have achieved the immediate distribution base had we not bought

Messel. It also gives us an enormous advantage on the UK domestic corporate finance side. Messel are brokers to over 80 companies, including many household names such as Glaxo, Harris Queensway and Trafalgar House. Our aim is to find ways in which we can offer these companies additional services from the resources of the group. In addition, we have brought in outside expertise; Bill Harrison, a former director of one of the leading UK merchant banks joined us earlier this year to lead our overall domestic investment banking efforts.' Harrison himself, having moved to Shearson Lehman, is quick to point out the potential synergy in combining the investment banking abilities of Wall Street with the corporate finance skills of London. 'I would be the first to recognise UK merchant banks have their own skills and you can't knock them', he comments, 'but the Americans have their own, too, and have engineered some highly ingenious products. Our aims are to develop inside Shearson Lehman the best of both cultures and this does not preclude us working together with UK merchant banks where this may benefit all parties.'

In 1981, American Express became a principal player on Wall Street when it acquired Shearson Loeb Rhoades, a large brokerage firm. Following this, their acquisition in May 1984 of Lehman Brothers, one of New York's oldest and most prestigious investment firms, created the second largest securities house on Wall Street and one of the three largest companies in the world. The combination gives Shearson Lehman Brothers an unmatched worldwide institutional and retail distribution network, to back its position as the biggest market maker on Nasdaq, the American computerised stockmarket that London is trying in many ways to emulate. This in itself gives the Shearson Lehman Messel group a clear advantage over purely British combinations.

'Shearson Lehman do more exclusive sales of US companies than any other bank in the United States and British merchant banks over there do not have that business. Similarly, American investment banks do not have the natural links with UK companies that Messel has and we believe we are uniquely well placed to match those sales of US companies with UK buyers', explains Giles Coode-Adams of Messel. 'Our research will reach a vast audience throughout the US through Shearson Lehman's sales network.' He cites the half a million shares of one well-known British company and client of Messel which were immediately snapped up after the announcement of a profits forecast which was relayed throughout the network of Shearson Lehman's salesmen as just one example of how the combination of client bases and trading experience can work so well.

The same kind of cross-fertilisation works on the corporate finance side, especially when it comes to the growing market for expertise in international mergers and acquisition business. While UK companies to date have been involved in more M & A business with US companies than with European companies, the weak dollar and political influences have more recently restricted US bids for British companies.

'We see great opportunities for British companies to internationalise their shareholder base', says Coode-Adams. 'Although equity is frequently a more expensive form of financing than debt, expansion can be limited if done with too much debt. There are

major advantages in broadening the shareholder base. For instance, a large percentage of the company's business may be overseas where the additional exposure of doing an international equity issue can give a company a tremendous marketing boost. Of course, it is vital that a broker maintains the interest of international investors to keep those benefits.

'At the moment most UK companies raise virtually all their money in the UK. I believe they should be seeking to ensure they can raise money on other international markets. Although funding on the domestic market may be easy enough now, this situation may not prevail much longer. Companies with a capital base in the region of £250 million to £1 billion in the UK should, we think, be taking this view from now on', says Coode-Adams. 'Five years down the line 50 per cent of their capital could easily be raised in non-domestic markets; whether it be in the form of equity, Eurobonds, convertible debt, ADRs (American Depositary Receipts) or money raised through the Japanese markets.'

To date, Shearson Lehman has acted as principal underwriter for two UK equity issues in which Messel acted as broker, including a sterling placement for the Hawley Group and as principal advisor to Allied Lyons in its takeover bid for Hiram Walker. 'We will shortly be at the stage where we can lead a major publicly contested take-over bid in the UK', says Bill Bardel, head of Corporate Finance at Shearson Lehman. 'We have a lot of experience in the US behind us as well as a specialist team in London. Our M & A team is headed by Michael Stallibras, who has extensive UK merchant bank corporate finance experience. He has been joined by a former colleague, Phillip Buscombe, and also Steven Berger from our New York M & A Department.'

'The trend towards greater risks and greater rewards will continue', says Bardel, 'and if you're going to compete in the market, the ability to commit capital will be mandatory. Earlier this year, we raised US$700 million for the World Bank in three different markets simultaneously without needing anyone else's help as manager or underwriter'. Coode-Adams agrees that the old approach to underwriting in the UK is on the wane. 'The legal

Above: Giles Coode-Adams, head of Corporate Finance at L. Messel. Right: Bill Bardel, head of Corporate Finance at Shearson Lehman Brothers International

restraints to the bought deal will be swept away with deregulation and change will be as inevitable as night follows day.' A bought deal often involves a single financial intermediary which is paid for executing, underwriting and distributing the entire issue, whereas in the case of a rights issue, first refusal on new shares is given to existing shareholders.

As Giles Coode-Adams points out, traditional pre-emptive rights and the more intangible factor of shareholder relations might in some cases discourage a company from using a bought deal simply to cut costs and that, 'in some cases it is right to give something to the underwriting institutions.' But if the bought deal method for equities becomes, as it has already for international debt issues, the most popular way of selling equity, the capital resources and distributive muscle of the lead-manager will be crucial.

Coode-Adams doesn't overlook the importance of options: 'We have arguably the best options team in London and believe effective hedging will be critical in a much wider range of situations. I see them becoming much more prevalent as a corporate finance tool. We'll also be dealing mainly net so our ability to quote a competitive net price in large amounts will be crucial. To do this you must have a good distribution base, a market-making talent and expertise in hedging. Come Big Bang, we will have all three.'

From October there will be 27 market makers competing in a new gilt-edged market where once there were only six major brokers and three jobbers. While net UK gilt sales will continue to grow, the pace of growth in government funding which has previously led gilts to dominate equities is unlikely to be sustained.

'Execution in the gilt market will become much more efficient, especially for large institutions', says Phillip Howard who directs overall trading of international fixed income securities from London. 'At the moment the larger investor has to pay high commissions on long gilt bargains and deal with a limited number of jobbers.' Howard believes that of the initial 27 market makers perhaps half a dozen will make money, another eight will be consistently involved and the remainder will become only marginal players by the end of 1987. 'We'll do well', he adds, 'because our approach is more integrated than most others and our resources will be specifically focused. It will not be an unfamiliar market to us. We are one of the top three houses trading in government securities in the US and Messel is one of the largest brokers in the UK.

'The different markets in stocks, bonds, futures, and foreign exchange are all becoming so integrated that we can no longer afford to focus on the profitability of one dealing area. Rather we have to consolidate our efforts, perhaps losing on one trading book to gain on another, so that we end up with an overall profit on a series of deals. As you can imagine, fast efficient information flows and communications between separate trading areas are vital. We've set ourselves up to be able to do this and that should give us a dealing edge.

'Fixed interest trading is an important part of Shearson Lehman's business worldwide and we see gilts market making becoming a substantial part of our UK domestic financial services.' The same way, Howard sees growth in other domestic fixed interest markets such as the sterling debenture market. When long-term fixed interest rates sink low enough to attract borrowers, he also anticipates a fixed interest mortgage-backed securities market opening here. Meanwhile, at the short end of the fixed interest market, Warren Spar, who heads up the commercial paper operations, adds, 'We also see immense scope in applying our well known US dollar CP capabilities to the fledgling sterling commercial paper market', pointing to the success of the £60 million issue for Allied Lyons in July which achieved a weighted average interest rate one basis point below London interbank bid rates which was part of the US$500 million multi-option facility jointly arranged by Shearson Lehman and Baring Brothers.

Commenting on the fixed interest market overall, Howard continues: 'Initially there'll be a lot of competition in terms of price-making and it probably won't be easy for the customer to decide who's the best firm to do business with. Our experience and ability to manage the business then becomes crucial. We're already developing up-to-the-minute research and technology to back it up and our experience in the US gives us a headstart. A lot of our emphasis is on understanding complex products that give our clients an edge.' Shearson Lehman Brothers International has developed a full product range in caps, futures, options and both interest and currency swaps over recent years. Price discrepancies in such esoteric instruments tend to be more pronounced than in others and so expertise in such areas brings its rewards.

'We are set for a fascinating time in the international capital markets', concludes Gelardin, 'where it is clear that the major Western governments including Japan (where we were granted a securities branch license earlier this year) are beginning to liberalise their markets; Switzerland, West Germany, France and Holland have followed suit. It will be extremely difficult for smaller financial operators to meet the demands of international companies in the way that we are able to do. In the UK we intend to build on the strength of the businesses which we're already familiar with and capitalise on the natural growth areas. We have the three prime requirements, as we see it, for Big Bang: the capital, the most advanced technology and the best people. It is up to us to "leverage" these exceptional resources in order to achieve our very ambitious goals.'

Acknowledgements

Many of the photographs in this book were provided by the individuals depicted or their organisations. Other credits are to the BBC Hulton Picture Library (pp. 25, 27, 28, 81), Camera Press (pp. 30, 115, 122), Central Press (p. 32), the Daily Telegraph Colour Library (p. 125), the Financial Times (pp. 31, 57, 117), Godfrey Argent (p. 114), the Guildhall Library (p. 91), Nick Hadfield (p. 52), Tom Hanley (pp. 68, 99, 116), Pat Hodgson (pp. 18, 22, 23, 88), the Image Bank (pp. 12, 127, 132, 136), Lewis Photos (p. 97), The Hon. Mrs. F. E. I. Mason (p. 94), Photographia (pp. 113, 123), the Stock Exchange (pp. 46, 102, 144), Tony Stone Worldwide (pp. 8, 140), UP International (p. 101). The charts were drawn by Richard Natkiel Associates. The book was designed by John White.